CAMBRIDGE COUNTY GEOGRAPHIES

General Editor: F. H. H. GUILLEMARD, M.A., M.D.

T0352090

WORCESTERSHIRE

Cambridge County Geographies

WORCESTERSHIRE

by

LEONARD J. WILLS, M.A., F.G.S.

FELLOW OF KING'S COLLEGE, CAMBRIDGE

With Maps, Diagrams and Illustrations

Cambridge :

at the University Press

1911

CAMBRIDGE UNIVERSITY PRESS
Cambridge, New York, Melbourne, Madrid, Cape Town,
Singapore, São Paulo, Delhi, Mexico City

Cambridge University Press
The Edinburgh Building, Cambridge CB2 8RU, UK

Published in the United States of America by Cambridge University Press, New York

www.cambridge.org
Information on this title: www.cambridge.org/9781107669680

First published 1911
First paperback edition 2013

A catalogue record for this publication is available from the British Library

ISBN 978-1-107-66968-0 Paperback

CONTENTS

ILLUSTRATIONS

ILLUSTRATIONS

The illustrations on pp. 17, 57, 136 are from photographs by
the author; that on p. 15 by H. W. Taunt and Co.; those on
pp. 34, 39, 40, 50, 96, 97, 104, 135 by Messrs Mallett and Sons;
those on pp. 58, 109 by Messrs Tilley and Son; that on p. 124 by
H. J. Whitlock and Sons, Ltd.; that on p. 146 by Messrs Valentine
and Son; that on p. 150 by J. Littlebury; those on pp. 128, 129,
131 by Mr Emery Walker; that on p. 64 is from a photograph
supplied by the Royal Porcelain Works; that on p. 65 from a
photograph supplied by Messrs Cadbury; those on pp. 54, 111
are from photographs by F. Gegg; that on p. 108 is by
J. Jacques, Jun.; that on p. 87 is reproduced by permission of
Messrs Cassell and Co., Ltd.; and those on pp. 6, 8, 10, 11, 21,
22, 23, 31, 47, 67, 72, 76, 78, 81, 89, 92, 93, 94, 99, 101, 102,
103, 106, 110, 114, 127, 137, 138, 140, 142, 143, 144, 145, 148,
151 are from photographs by Messrs F. Frith and Co., Ltd.

1. County and Shire. The word Worcester: its Origin and Meaning.

At the very beginning of our study of the geography of our county we are forced to recognise the intimate connection of that science with history. History and geography are in fact sister studies and neither can be truly grasped without some knowledge of the other. Thus at the outset we are obliged to grope in the obscure annals of the past for the meaning of the name and for the origin of the county; while later it will be apparent that the geographical position of the shire has affected its history far more deeply than we could easily imagine. It is important to keep this relationship well to the front and to endeavour to recognise that the present-day features of landscape, agriculture, and industry are far from being accidental, and that we must look for explanations of them in the past history, geological and human, of the county.

If we allow our minds to travel back about 1500 years or so to the time of the early English settlement, we find England not a united whole but composed of numerous petty kingdoms. These divisions were to some extent tribal a fact which must be borne in mind when we seek for the origin of our present political divisions known as counties or shires. When we find a name, like Sussex or Kent, obviously complete in itself, we may rest assured that it denotes one of these ancient kingdoms; while a county-name ending in *shire* shows equally well that the division bearing such a name originally formed a *shire* or *share* of a larger kingdom—the portion that has been *shorn* from it. In the case of Worcestershire, the share was part of the large Midland kingdom of Mercia, which extended from the Humber to the borders of Wessex in Oxfordshire and Berkshire.

As these small kingdoms gradually became merged under the rule of one king of England, Earls were appointed by the Crown to govern either the whole or part of each. These corresponded, roughly speaking, to our present Lord-lieutenants of counties. Thus we come to recognise that the present-day political divisions and administrators are no new invention, and that the mode of government instituted by our Saxon forefathers 1as until quite recently proved itself efficient.

We may perhaps ask why the word *county* has replaced the original term *Earldom*. The use of the former word must be ascribed to the introduction of French by the Normans as the official language. It is derived from the French *comté* which implies the dominion of a count. The great lords were however always known as Earls in England, but the French form—Countess—still survives in the title of their wives.

Of the very remote history of Worcestershire we know but little. There are no written records of the district before the Anglo-Saxon settlement, which took place at a comparatively late date in this part of England. We know that the tribe called the Hwiccas occupied a part of the district which is now embraced by the counties of Warwickshire, Worcestershire, and Gloucestershire. They were at first under the sovereignty of Wessex but about A.D. 628 were conquered by Penda, who added their territory to Mercia. That the tribe was of importance we know from the fact that the early bishops of Worcester were ordained as bishops of the Hwiccas, the diocese in those days being roughly co-extensive with the tribal boundaries. In fact it was not until Henry VIII's time that the see of Worcester was curtailed to something like its more recent dimensions by the institution of the diocese of Gloucester.

Now although it is well established in what way the diocese came into existence, there is considerable doubt as to who instituted the political county. The present division of England into counties is often attributed to King Alfred. This part of England appears to have been divided into shires after its reconquest from the Danes by Edward the Elder in 922.

Now let us consider the origin of the word Worcester. Here again we must go back to very early times, and in doing so we are met by even greater difficulties. This district appears to have been once densely clothed by forests and only sparsely populated. Accordingly very little is known of its history. There is said to have been

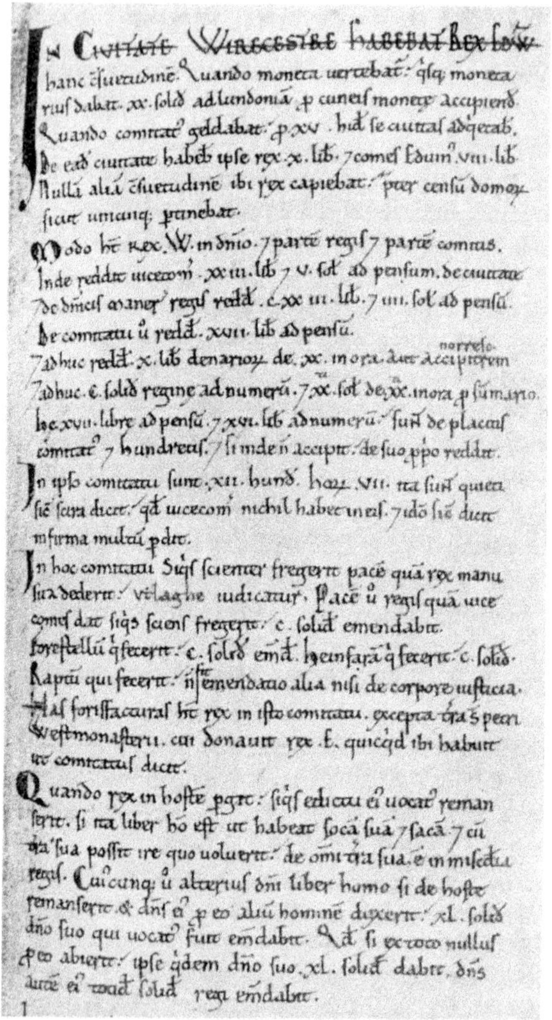

A Page from Domesday Book showing an entry under Worcester

a British settlement at Worcester, but there is no proof that its name was Caer Wrangon as some people have tried to maintain.

In Roman records no definite mention of Worcester has been found, but the Latinised word Vigorna or Wigornaceaster of the old English chronicles may well be taken to represent its Roman name. It is from this word that the present name appears to have been derived, and it was the custom of the late bishop of Worcester to sign himself Wigorn (*Wigornensis*).

The name of the county in the Domesday Book appears as Wirecestrescire, which approaches very closely to the sound which the name still possesses in the broad speech of the country folk. But, although we more or less know the history of the word Worcester, its meaning and origin still remain a secret.

2. General Characteristics.

Ever since our English forefathers drove the Britons into Wales, the natural position of Worcestershire has made it virtually a border county against the Welsh. It came, for instance, under the jurisdiction of the Court of the Welsh Marches. The county has accordingly been the theatre of many troublous scenes; more so perhaps than other border counties, on account of its pronounced loyalty to the Crown, which has won for the city of Worcester the motto *Civitas in bello, in pace fidelis*. These Royalist tendencies led the county to take a prominent

The Severn Valley from the Malvern Hills

part during the time of the Civil Wars, which culminated in the battle of Worcester in 1651.

That in early times the hill country on the western side of the county was of the greatest strategic importance may be gathered from the line of earthworks which cap the more prominent summits. These hills probably belonged at times to the Britons, at others to the English, while the line of the river Severn appears to have constituted the real boundary between the two peoples, for at that time the marshes of the Severn, up which the tide swirled perhaps as far as Bewdley, formed an almost impenetrable barrier dividing the two districts.

A second feature which must have exercised a powerful influence over the history of the county was the great extent of forest land, part persisting until a relatively recent date. In mediaeval times a separate code of laws was maintained in them. The forests are now shrunk to small dimensions, and Worcestershire can no longer be considered a forest-country, but in the venerable oaks and elms which stand in the hedgerows and fields we find some evidence of its former glory. The observer can still see from the top of the Lickey Hills, or from the Worcestershire Beacon, a wonderful wooded landscape, though now broken up by rich pastures, golden cornfields, or fruitful orchards.

One of the features of Worcestershire which strikes the eye is the prevalence throughout large areas of a dark red soil. This is due to the red rocks which underlie the greater part of the county. From them the houses, both of brick and of stone, draw the warm colours which contrast so strongly with the grey monotony of many districts.

Worcestershire may in a rough way be divided into two areas, given over respectively to manufacture and to agriculture. The latter claims the larger part, and it is only in the north-east of the county that the manu-

On the Severn—Forest-land near Bewdley

facturing district of Birmingham and the Black Country has overflowed into it.

The county has always been noted for its fruit, so much so, in fact, that the pear forms part of the coat-of-arms of the city of Worcester. The whole of the Vale of Evesham and Pershore is one great fruit and market garden, while almost throughout the county apples and pears are grown to make cider and perry.

But in places the ground rises into hills and tracts of upland country where wiry grass, heather, gorse, and bilberry flourish ; yet we must not forget that the county owes its fertility, in part at any rate, to its proximity to these hills and those of Wales, from which the streams throughout the ages have been bringing down the rich alluvial mud and loam to form the valley lowlands.

The natural fertility of the soil in the vales of Severn and Avon is increased by the equable climate which the county enjoys. These factors, together with a ready access to good markets by means of railways, place the county in a position of peculiar advantage from an agricultural point of view.

3. Size. Shape. Boundaries.

Worcestershire is an entirely inland county, although the Severn provides ready access to the sea, for the river is navigable throughout almost its whole course through the county, and vessels of 200 tons can come up to the city of Worcester.

The county is of roughly quadrilateral shape but of extraordinarily irregular outline, and it may well be claimed that it is first among counties in its almost complete disregard of natural boundaries. The ridge of the Malvern Hills in part and a few short stretches of river form the only natural delimitations. This, at first sight, seems remarkable, since it would be expected that the lines of the Severn, Teme, and Avon would have

Worcester Cathedral from the Severn

been chosen as the natural limits. In seeking for an explanation of this we must remember that throughout the history of Worcestershire the Church has played an influential part. It was the Benedictine monks who first cultivated the land to any great extent, and naturally they chose the valleys in which to start their labours.

The Ivy Scar Rock, Malvern

As they maintained possession of their lands after the Conquest, we probably see in the county boundary remains of the limits of the church lands of Worcester, Evesham, Pershore, and Malvern. At the same time we may note that many of the early tribal boundaries were formed rather by the watersheds than by the river courses, for the invaders pushed their way up the rivers and seized the land on either side. It is therefore possible that the same causes took effect in this part of Britain.

In point of size Worcestershire comes among the smaller counties of England. It embraces an area about one sixty-seventh of that of England, and comprises 480,560 acres or 751 square miles. This is the acreage of the ancient county. That of the present administrative county is 473,328 acres, since a portion of the ancient county—the boroughs of Worcester and Dudley—is now excluded from the administration. The county measures as much as 45 miles across in several directions and has a circumference of about 220 miles.

Worcestershire is bounded on the east by Warwickshire, on the south by Gloucestershire, and on the west by Herefordshire; while to the north-west and north respectively are the counties of Shropshire and Staffordshire.

We have already noticed the peculiarly irregular outline of the shire, but even more curious than this are the numerous detached portions which lie inside the limits of other counties. These two features are probably due to the same cause. The detached portions of Worcestershire are as follows :—(i) the important town of Dudley, which forms a kind of island of Worcestershire in Staffordshire, (ii) the parish of Shipston-on-Stour in Warwickshire, (iii) the parishes of Evenlode, Cutsdean, Blockley, and Daylesford in Gloucestershire. There were formerly other detached parishes belonging to the county which have been exchanged for portions of other shires lying within the boundary of Worcestershire.

In the case of Worcestershire an explanation of this

lies in the fact that all the parishes mentioned above, except that of Dudley, formed part of the Hundred of Oswaldslow. This Hundred contained by Charter of Eadgar all the lands of the Church and Monastery at Worcester, where St Oswald was then bishop. Its court was held at Worcester and accordingly the several parishes which composed it came to form part of the county. The reason for the inclusion of Dudley is of a similar nature. It formed part of the domain of the Norman Fitzarnulf and was the only castle in Worcestershire mentioned in Domesday Book. So we see that the actual present-day boundaries of the county are of very great antiquity in these cases.

It has now been settled by Act of Parliament that these outlying portions may be joined with the district in which they are situated, if both parties interested in the locality agree to the amalgamation.

4.   Surface and General Features.

It will be seen later that the geological structure and the nature of the rocks composing the surface of the county are very varied ; and, partly as a result of this and partly on account of the great differences in altitude, the physical features of Worcestershire are very diversified.

In dealing with the general topography it will be convenient to divide the county into three areas which are markedly distinct, and characterised by particular types of scenery. These we may call the north-eastern upland district, the central area embracing also the vales of Severn and Avon, and the western upland district. If one takes up a position on the top of the Clent or of the Lickey Hills, the meaning of this three-fold division is apparent.

Let us suppose ourselves to be on the Lickey Beacon Hill, on one of those glorious days after rain when the west wind blows strongly and the distance is outlined in darkest blue. To the north and immediately around us lie the north-eastern uplands, with the Clent Hills and the various ridges of the Lickeys on which we stand. Farther to the north the land is still high and more or less covered by manufacturing towns such as Dudley, Rowley, Cradley, and Oldbury, that have risen here from the wealth of mineral which underlies them. Nearer to us is the ridge of Frankley, with its tuft of beeches and high ground stretching away to Birmingham. South by east the Clent-Lickey range tails out into a narrow ridge that runs past Redditch ; while farther to the east is a long flat-topped escarpment forming the edge of Warwickshire.

But if we turn to the south we see a grand expanse of well-wooded land, sloping away from our feet, and stretching almost as far as the eye can see to another long flat-topped escarpment, that of the Gloucestershire Cotswolds, while Bredon Hill stands out as an advanced post thrust forward from them into Worcestershire. In reality it is this escarpment which one thinks of as the southern limit of the shire though, as we have already seen, the

Bredon Hill

political boundary pays no respect to such features. Farther to the west there meets us an outline savouring more of mountains and of Wales; first the glorious profile of the Malvern Hills, the most essential part of many a Worcestershire landscape ; then, north of them, the Abberley Hills with the characteristic clump of pines on Woodbury Camp, while farther again to the north rise the Clee Hills and the high ground to the west of Bewdley. All the broad distance between us and this circle of hills we may call the central area. When we come to examine it more closely, we find here also a far greater variety of topography than would be imagined from our bird's-eye view on the Lickey. On the south runs the Warwickshire Avon, in a flat, rich valley of gardens and orchards ; on the west is the quickly-flowing Severn, in a narrow trough which only broadens out below Worcester itself; while between these two vales is a plateau-like country of orchard, woodland, and plough. We must not, however, forget Bredon, that weather-glass of the Vale of Evesham, of which the rhyme runs thus:

> " When Bredon Hill puts on his hat,
> Ye men of the Vale beware of that,
> When Bredon Hill doth clear appear
> Ye men of the Vale have nought to fear."

This is the one conspicuous feature of the central area and forms the background to many views. It is, as we have said, a foretaste, both in appearance and structure, of the Cotswolds, with its long southerly and steep northern slopes. Much of the central area was formerly

A Worcestershire Lane

forest land. Malvern Chase extended from the Severn to the top of the Malvern Hills and from the Forest of Dean to the Forest of Wyre, while the Forest of Feckenham spread over a great part of the plateau mentioned above.

The western district is comprised in the long arm of Worcestershire which stretches up the Teme Valley between Shropshire and Hereford. The Malvern, Martley, and Abberley line of hills and Wyre Forest are its eastern limit. This is a country of rolling uplands and deep valleys, rich in orchards and hop-gardens, and drained by the beautiful Teme, that quick-flowing river that breaks its way through the sentinel hills at Knightsford Bridge. Malvern is preeminently a Worcestershire mountain—for mountain we may well term it by reason of its majestic outline and the appearance of height conveyed by the abruptness of its rise out of the plain of the Severn. Abberley is another fine group of hills, and these two, just as the Clent and Lickeys in the north-eastern uplands and Bredon in the central area, stand out as characteristic of the western district. Here too we have one of the largest remnants of oak forest in the kingdom. The Forest of Wyre, some 8000 or 10,000 acres in extent and composed almost exclusively of oak-scrub, are all that remains of a greater woodland which has mostly disappeared, no doubt to provide fuel for the iron furnaces in the days before coal was used.

5. Watersheds and Rivers.

The physical geography of Worcestershire is considerably simplified by the fact that the county lies almost completely in the drainage area of the Severn and its tributaries, of which the most considerable are the Stour, the Teme, and the Warwickshire Avon.

The great central watershed of England runs, however, through the extreme eastern part of the county, where the headwaters of the Cole and Rea, feeders of the Trent, are to be found, the former on the high ground to the east and south-east of King's Norton, and the latter near Frankley and Northfield.

The rest of Worcestershire is in the Severn watershed. This river, "the greatest water ornament and prodigal benefactor of our county," as the Worcestershire historian Habington puts it, enters the county in the extreme north at Upper Arley, and flows throughout its length with swift brown current almost due north and south until it leaves it about a mile north of Tewkesbury. If we look at the geological map at the end of this book, we notice that its course runs entirely on the soft Triassic rocks.

Let us trace the river from its entrance to its exit. Above Bewdley, steep walls of wooded hills shut in the Severn on both sides, while a mile below the town the precipitous boss of rock known as the Blackstone Rock and the slopes above Ribbesford afford a quite Welsh type of scenery. To the west of this reach lies the Wyre Forest and the high country round Rock.

2—2

Bewdley, once a much more important place, commanding a ford and a main trade route to Wales, is now a quiet though dignified little town, and very different to Stourport, which is situated a few miles lower down. Here the river Stour, an unclean and sluggish stream, the Staffordshire and Worcestershire Canal, and the Stourbridge Canal empty themselves into the Severn. It is said that formerly the tide ran up nearly as far as Bewdley. Be this as it may, the Severn certainly formed an important means of transport before the days of railways. At first boats came up to Bewdley, which was the natural outlet for the Black Country trade, but on the opening of the canals much of the traffic was diverted to them. These again lost their importance on the introduction of railways, and looking at them nowadays it is difficult to realise that it was these canals which brought Stourport into existence.

Due west of Stourport is Arley Kings, the home of Layamon, one of our earliest English poets. The church, in which he is commemorated, occupies a commanding position overlooking the busy town opposite.

As we follow down stream, still in a deep and fairly narrow trough, we pass the great wood of Shrawley with its peculiar small-leaved lime trees, the lock and bridge at Holt, and then on the left bank the river Salwarpe, whose very name smells of the salt of Droitwich. Here too is the end of the short Droitwich Canal, which completes water communication between that town and the sea. And so we arrive at Worcester, with its fine cathedral perched on a commanding height on the left bank of the

river. We must not however pause here, but crossing
the river by the stone bridge let us make for Powick,
famous for the battles fought here in the Civil War.
Near Powick the Teme joins the Severn. It is a stream
which savours of the West and of Wales and makes a
glorious contrast to the other rivers of Worcestershire.
The Severn, though swift flowing, is always muddy, and

The Junction of the Teme and the Severn near Powick

it is said that in its waters the salmon never rises to the
fly, whereas the Teme is preeminently a trout and grayling
river, swift and clear. It enters the county at Tenbury,
some 18 miles from Powick as the crow flies, and provides some of the most beautiful scenery in the county.
Especially noteworthy is the reach near Knightsford
Bridge, where the stream breaks through the line of

hills which are the real boundary of Wales, and where its valley broadens out to merge into the Vale of Severn. There are no important towns on the Teme, though its valley is rich in orchards and hop-gardens.

From Powick to the county boundary at Tewkesbury the Severn runs through a more open valley, past the ancient ford, "The Rhydd," past Severn Stoke and Upton,

The Avon at Fladbury

still flowing swiftly, until near the red marl cliff of the Mythe Tute it passes into Gloucestershire.

At Tewkesbury, though not actually in the county of Worcester, the Severn receives the waters of the Avon. This delightful meandering stream lies in the wide vale of Evesham, entering the county near Cleeve Prior, some 16 miles in a direct line from Tewkesbury. Between these two places, the river bends and winds for some six and twenty miles round the northern side of Bredon Hill. Evesham and Pershore are the important towns on this stretch, though there are many villages and several mills, all of which are picturesque and worth a visit.

The Avon was formerly navigable as far as Stratford, but the locks between that town and Evesham are now out of repair. Between the latter place and Tewkesbury,

The Avon near Evesham

however, steamers ply in the summer, but the river is not used for commercial purposes.

The Avon differs from the Severn in many ways, but one curious point is perhaps worth special notice. This is that salmon and lampreys never run up the Avon though they abound in the Severn. In both streams, however, coarse fish are abundant.

Worcestershire possesses no natural lakes, but there are a few large artificial reservoirs, such as those at Frankley and Cofton Hackett near Barnt Green, and large sheets of ornamental water at Westwood Park near Droitwich, and at Hewell Grange.

6. Geology and Soil. (*a*) General Considerations.

By Geology we mean the study of the rocks, and we must at the outset explain that the term *rock* is used by the geologist without any reference to the hardness or compactness of the material to which the name is applied ; thus he speaks of loose sand as a rock equally with a hard substance like granite.

If we examine the water of any river we find that it is loaded with fine particles of mud and sometimes with sand or pebbles. Further it carries some lime in solution. Now it has been found that these materials are deposited in lakes or in the sea, and subsequently become rocks. When the material is coarse, it forms gravels or conglomerates; when finer, sandstones or clays. Limestones are produced from the carbonate of lime that is dissolved in natural waters and afterwards deposited again, and also from the accumulation of shells. Sometimes sand blown by the wind accumulates as dunes, which may harden into sandstone. In nearly every case the material is deposited in almost horizontal sheets, beds, or strata, and rocks thus formed are known as sedimentary rocks.

It is evident that of any two such beds lying one on the other, the upper one is the newer. Although originally flat the beds may afterwards be tilted or folded as a result of movements in the earth-crust (see diagrams). As soon as any part of the strata comes above sea-level streams start to flow on its surface, and it gets gradually eaten away by the action of the sea, rivers, or glaciers. This action, tending to wear the folds down to an even surface, is known as Denudation. If the rocks sink once

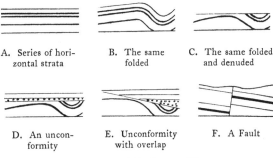

A. Series of horizontal strata

B. The same folded

C. The same folded and denuded

D. An unconformity

E. Unconformity with overlap

F. A Fault

Diagram to explain Strata Changes

more below the sea, the next bed deposited will lie horizontally on the worn-down edges of the older one. D in the annexed diagram is a section to show such an "unconformity," as it is termed.

A variety of changes may occur in a series of beds after deposition. Thus they may become hardened and consolidated from a non-coherent state, or they may become cracked by "joints." These cracks are important both in their effect on scenery and also in their practical use to the quarryman, who avails himself of their presence as the easiest method of extraction of the rock. Again, as a result of great pressure applied sideways, rocks may be so changed that they may be split into thin slabs, which usually, though not necessarily, split along planes standing at right angles to the horizontal. Rocks affected in this way are known as slates.

Sometimes the beds have been broken across vertically, and those on one side of the crack have moved either downwards or upwards relatively to those on the other. Such a crack is known as a *fault*. These are not often visible at the surface, since the upraised part gets worn away by rivers and other agents of denudation. They are, however, sometimes of grave commercial importance; for example, there is in Worcestershire a large fault which throws down the coal-bearing beds from the surface to great depths.

What can we learn from a rock about the history of the earth ? First we can sometimes gather from the nature of the deposit under what conditions it was formed. Thus coarse sandstone cannot have been laid down far from the shore, nor pure limestone near to it, but out in the clear sea, where muddy rivers did not reach. Secondly, we can argue from the presence of unconformities that our land has at times been below and at other times above the sea-level. We can thus to some extent reconstruct the geography of past ages. Thirdly, we can tell the relative age of the beds; for of any two the lower is the older. Now in sedimentary rocks we often find the fossil remains of animals and plants which were entombed during their formation. Beds of a given age usually contain the fossils peculiar to that age, so that we can generally assign an age to a fossiliferous rock without necessarily seeing what lies above or below it. Fossils are further of great assistance in interpreting the conditions under which the rock was deposited.

There remain rocks of another type known as *Igneous*. They include such rocks as granites, diorites, and basalts, which have previously existed in a molten state in the hot interior of the earth. When still in this condition they have been shot out or poured forth as lavas from volcanoes, or have been forced into other rocks and cooled in the cracks and other places of weakness in the earth-crust. Igneous rocks are, however, comparatively infrequent in this district, and so at present we need say no more about them.

If we could flatten out all the strata of England, and arrange them one over the other and bore a shaft through them, we should see them on the sides of the shaft, the newest appearing at the top and the oldest at the bottom, as shown in the table (p. 28). Such a shaft would have a depth of between 100,000 and 150,000 feet. The strata or beds are divided into three great groups called Palaeozoic or Primary, Mesozoic or Secondary, and Cainozoic or Tertiary, and below the Primary rocks are the oldest rocks of Britain, which form as it were the foundation stones on which the other formations rest. These may be spoken of as the Pre-Cambrian rocks. The three great groups are divided into minor divisions known as systems. The names of these systems are arranged in order in the

	Names of Systems	Subdivisions	Characters of Rocks
TERTIARY	**Recent Pleistocene**	Metal Age Deposits Neolithic ,, Palaeolithic ,, Glacial ,,	Superficial Deposits
	Pliocene	Cromer Series Weybourne Crag Chillesford and Norwich Crags Red and Walton Crags Coralline Crag	Sands chiefly
	Miocene	Absent from Britain	
	Eocene	Fluviomarine Beds of Hampshire Bagshot Beds London Clay Oldhaven Beds, Woolwich and Reading Thanet Sands [Groups	Clays and Sands chiefly
SECONDARY	**Cretaceous**	Chalk Upper Greensand and Gault Lower Greensand Weald Clay Hastings Sands	Chalk at top Sandstones and Clays below
	Jurassic	Purbeck Beds Portland Beds Kimmeridge Clay Corallian Beds Oxford Clay and Kellaways Rock Cornbrash Forest Marble Great Oolite with Stonesfield Slate Inferior Oolite Lias—Upper, Middle, and Lower	Shales, Sandstones and Oolitic Limestones
	Triassic	Rhaetic Keuper Marls Keuper Sandstone Upper Bunter Sandstone Bunter Pebble Beds Lower Bunter Sandstone	Red Sandstones and Marls, Gypsum and Salt
PRIMARY	**Permian**	Magnesian Limestone and Sandstone Marl Slate Lower Permian Sandstone	Red Sandstones and Magnesian Limestone
	Carboniferous	Coal Measures Millstone Grit Mountain Limestone Basal Carboniferous Rocks	Sandstones, Shales and Coals at top Sandstones in middle Limestone and Shales below
	Devonian	Upper Middle } Devonian and Old Red Sand- Lower } stone	Red Sandstones, Shales, Slates and Lime- stones
	Silurian	Ludlow Beds Wenlock Beds Llandovery Beds	Sandstones, Shales and Thin Limestones
	Ordovician	Caradoc Beds Llandeilo Beds Arenig Beds	Shales, Slates, Sandstones and Thin Limestones
	Cambrian	Tremadoc Slates Lingula Flags Menevian Beds Harlech Grits and Llanberis Slates	Slates and Sandstones
	Pre-Cambrian	No definite classification yet made	Sandstones, Slates and Volcanic Rocks

accompanying Table. On the right hand side, the general characters of the rocks of each system are stated.

The geology of a district affects its history in every way; for on it depends the nature of the soil, the presence or absence of mining industries, the quality and abundance of the water, and above all the natural scenery. It is proposed to attempt to show in the sequel how the great variety of scenery and industries that we find in Worcestershire depends on the geological structure. To help in understanding this and to make the following remarks on the geology more intelligible, the reader is referred to the Geological map at the end of the book, which is coloured to show the ages of the rocks present at the surface.

It will be seen on comparing the topographical with the geological map that there is a marked coincidence between the hill country on the western side of the Malvern, Abberley, Wyre Forest line, which we have termed the western uplands, and the Palaeozoic rocks. The central area on the other hand is cut out of the soft Mesozoic rocks, whereas the rising ground to the north-east is again an indication of the presence of older formations. The generalised section across the northern part of the county from the north-east to the south-west (p. 32), shows the relationship of these three areas. The western uplands are composed of gently-folded strata, except on their eastern extremity, where the beds have been much disturbed. Then comes a fault (F^1) which has let down all the Mesozoic rocks to the east into a sort of trough, the further side of which is also marked by a fault, F^2. The Mesozoic rocks between these two faults are depressed into a kind of basin, whose centre is about where the town of Droitwich stands. The eastern Palaeozoic area is much cut up by faults, and the beds are slightly folded.

7. Geology and Soil. (*b*) Rocks found in Worcestershire.

The oldest rocks in Worcestershire belong to the Pre-Cambrian formation which has been mentioned above as constituting the old foundation on which all the other rocks rest. The Malvern Hills owe their shape largely to the presence of these Pre-Cambrian rocks which are also to be seen at Barnt Green on the slopes of the Lickey Hills. In both cases they are igneous rocks and are succeeded unconformably by the Cambrian strata. The Pre-Cambrian syenite of Malvern is quarried chiefly for road metal. The Rednal and Bilberry Hills in the Lickey area are composed of Cambrian quartzites which are also quarried for road stone.

These two outcrops of Cambrian and Pre-Cambrian rocks at Malvern and at the Lickey Hills are surrounded by faults and are remarkable for the steepness of outline that the hills possess. During the Ordovician period, which succeeded the Cambrian, Worcestershire appears to have been land. The area was subjected to denudation, which removed part of the Cambrian and Pre-Cambrian strata and reduced the landscape almost to a plain over which

The Wych Cutting, Malvern Hills
(*In Pre-Cambrian Syenite*)

the Silurian sea spread. This brought with it the coarse sand which has now been consolidated into the Llandovery or May Hill Sandstone. The fossils in this rock are quite different from those of the Cambrian, but in many instances they resemble those of the next succeeding period which is known as the Wenlock Series.

These rocks are found extensively in the western Palaeozoic tract and as isolated patches surrounded by faults in the eastern area. Though largely composed of shales yet they are most famous on account of the wonder-

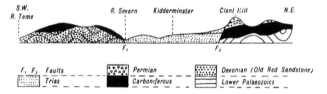

S.W. R. Teme R. Severn Kidderminster Clent Hill N.E.

F₁ F₂

F, F₂ Faults
Trias
Permian
Carboniferous
Devonian (Old Red Sandstone)
Lower Palaeozoics

Generalised Section across the Northern part of
Worcestershire

ful preservation of the fossils in some of the limestone bands. These limestones usually stand out as prominent ridges of which the Wren's Nest[1] near Dudley is an excellent example.

The Ludlow Series which succeeds the Wenlock is also composed of shales and occasional limestones which are likewise very fossiliferous, in marked contrast to the overlying Devonian rocks. These are singularly devoid of fossils and are made up of sandstones and shales known in this part of England as the Old Red Sandstone Forma-

[1] Just in Staffordshire.

tion. After the deposition of the Devonian rocks, Worcestershire again became part of a land surface which sloped towards the north into a basin-like area, which later became covered by a shallow lagoon, usually of fresh, but occasionally of salt water in which the coal measures of the South Staffordshire and Wyre Forest coalfields were laid down quite unconformably to the Silurian and Devonian rocks below.

Of these two, the Wyre Forest coalfield is only slightly productive, while a small part alone of the productive measures of the South Staffordshire field lies in Worcestershire, though unproductive Upper Coal Measures occupy parts of the north-eastern area as far south as Rubery. The productive measures near Dudley contain the famous "Thick" or "10-yard coal," a seam which is here 30 feet in thickness.

Following the Upper Coal Measures comes the Permian Formation. The deposits of this age rest with a slight unconformity on all below, showing that once again this district had for a short time become dry land. The nature of the Permian rocks is remarkable. They consist chiefly of what is known as a *breccia*, a rock composed of angular, unrolled fragments set in a fine matrix. These fragments are mostly of volcanic rocks, while the matrix is a red marl. It is thought that the blocks were broken off by frost on high ground and that the climate was generally dry, though occasionally subject to torrential rains which washed them down quickly without producing a waterworn appearance. We must accordingly consider Worcestershire at this time as under desert conditions, which were probably comparable to those of Central Asia of to-day. Without exception these Permian beds stand out as hills, and the main ridge of the Lickey and Clent range is formed of them.

The Triassic rocks which occupy the greater part of

Rock Dwellings near Blakeshall

the county may well have been formed under somewhat similar conditions. The Lower Trias or Bunter is characterised by red sandstones with bands of waterworn pebbles in the middle of the series. The lower sandstones of the Bunter have in several places been cut out into rock dwellings, some of which are still inhabited.   Much of the county underlain by the Bunter is given over to gorse commons and woods.

The Upper Trias or Keuper Formation consists of sandstones below and red marls above.   The sandstones of this group have been as a rule the formation chosen unconsciously by the early settlers on account of the good drainage that a porous rock of this nature ensures and because they are generally water-bearing.   Good building stone can be also obtained from them but at present they are only worked for this purpose at Bromsgrove and Inkberrow.

The Keuper Marls are red and green clays, which near Droitwich and Stoke Prior contain important beds of salt. This indicates that the lake in which they were formed was completely dried up at one time, and accordingly we may suspect that very dry conditions prevailed here then. And now the reader will appreciate the statement made above that the geology affects the history of the district. For the oldest industry and the one most affecting the life and history of the county is salt-making, which we owe to the climatic conditions prevailing in Triassic times.

Jurassic rocks are only found in the south of the county, but they are of considerable importance on account of the soil they produce and the vegetation they support.   The most important rock is the Lias Clay, which is a heavy blue clay underlying a triangular patch of country between Droitwich, Tewkesbury, and Evesham. The fossils in this are numerous and show by their nature that the sea had now spread over this area.

The oolitic limestones of Middle Jurassic age are unimportant in Worcestershire, though of interest. For it is the hard covering which they provide in places over the soft Lias Clay that produces the escarpment of the Cotswold Hills and the isolated hill of Bredon. The outlying detached portions of Worcestershire are mostly situated on such limestones, whose presence indicates a clearer sea than that of the Lias Clay. But it is notable that though clear it was shallow enough to allow of the formation of oolite. This is a rock made up of small round grains which resemble the roe of a fish, and it is from this resemblance that it obtains the name.

The oolitic limestones are the newest consolidated rock in Worcestershire. There are however various surface deposits, of which the alluvium of the river valleys is of great importance on account of its influence on agriculture. But of greater scientific interest is the Boulder Clay. This is a clay with stones in it which have not necessarily been derived from the rocks of the neighbourhood, but may have been brought by glaciers from distant sources. Such stones are known as *boulders*. They are often angular, grooved, and scratched. The glacier which brought the Boulder Clay of Worcestershire appears to have come from North Wales, bringing with it fragments of the Welsh rocks, some of which attain a diameter of two or three feet. The epoch at which this glacier existed is known as the Glacial Period and probably ended about the time that man appeared in this part of the world.

Glacial gravel often caps the hills, and round the

slopes of the Lickeys this produces a peculiar type of scenery. The gravel protects the tops of the hills while the unprotected slopes of soft sandstone are more quickly removed, and peculiar rounded outlines are produced.

The Boulder Clay is employed at Blackwell to make bricks.

8. Natural History.

Various facts, which can only be shortly mentioned here, go to show that the British Isles have not existed as such, and separated from the continent, for any great length of geological time. Around our coasts, for instance, are in several places remains of forests now sunk beneath the sea, and only to be seen at extreme low water. Between England and the continent the sea is very shallow, but a little west of Ireland we soon come to very deep soundings. Great Britain and Ireland were thus once part of the European continent, and are examples of what geologists call recent continental islands. But we also have no less certain proof that at some anterior period they were almost entirely submerged. The fauna and flora thus being destroyed, the land would have to be re-stocked with animals and plants from the continent when union took place, the influx of course coming from the east and south. As, however, it was not long before separation again occurred, not all the continental species could establish themselves. We should thus expect to find that the parts in the neighbourhood of the continent were richer in species and those farthest off poorer, and this proves to be the case both with plants and animals. While Britain has fewer species than France or Belgium, Ireland has still less than Britain.

It must not, however, be imagined that the same progressive diminution can be seen in the number of species as we travel from east to west in England. By far the greater number of common plants and animals are found over the whole of the country and their presence or absence depends chiefly on the nature of the habitat. This applies especially to plants, and as we pass on to a consideration of the natural history of Worcestershire we find that the great variety displayed in its geological structure and topography is reflected in the wealth and diversity of its flora, but to a far less degree in that of its fauna. We may perhaps be able to note some connection between the nature of the fauna and the flora associated with it.

On the barren slopes of the Malvern and Clent Hills the short wiry grass is in marked contrast with the lush meadows of the vales of Evesham and Severn. The Lickey Hills again are in places deep in bracken, heather, and bilberry, and in autumn red with mountain-ash berries, while the gorse-covered commons of the Bunter sandstone country round Kidderminster and Stourport are equally distinctive.

The country is on the whole so well cultivated that few wild spots remain, and for the most part the deep hedge banks are the nearest approach to wild nature that is possible. The marshes on the Severn below Worcester are now all drained, but within historic times the tide reached Longdon Marsh, which it is of interest to note was until recently rich in salt-marsh plants.

We may notice that the flora is mostly of a lowland

Clipped Yew Trees—Manor House, Cleeve Prior

type, but it may be worth while to see how this varies according to the nature of the soil. This is however an intricate subject and we can only note one or two peculiarities. The clematis known as " Traveller's Joy " is a very typical limestone-loving plant and in the south-eastern part of the county this charming creeper luxuriates over the hedges, a result of the soil which Liassic lime-stones provide. Again, on sandstones, gorse and broom are especially abundant, and the smooth arabis and mountain groundsel (*Senecio sylvatica*) are characteristic.

Worcestershire is a county conspicuous for its timber.

Mitre Oak, near Hartlebury

In the olden days it was thickly forest-clad, but most of the forest was consumed by the salt-pans and iron furnaces which until the middle of the seventeenth century used wood fuel ; and so the county gradually became what it now is—well-timbered, but without large stretches of woodland, with the exception of the Wyre Forest, where some 8000 to 10,000 acres of scrub oak has survived to the present day.

Hartlebury, the summer palace of the bishops, is famous for its elms, and there are several celebrated oak trees in the county. Of these the "Mitre Oak" at Hartlebury claims to be the descendant of the tree beneath which St Augustine held his famous conference with the Welsh bishops about the year 602.

In many cases it is more than probable that the old trees standing in the hedges are remnants of the forests which were allowed to remain as landmarks. They form quite a conspicuous feature in Worcestershire landscapes, as do also the fruit trees which until recently were often planted in the hedges. A peculiar plant which cannot fail to be noticed is the mistletoe, which grows in tufts in every orchard. In Shrawley woods on the banks of the Severn, near Holt Fleet, there is a heronry, but the wood is famous rather as the natural home of the small-leaved lime tree which is indigenous here.

Worcestershire has the usual quota of large animals. The stoat, weasel, fox, badger, squirrel, otter, mole, and various species of rats, mice, and bats are common. Deer are now only found in parks, though once upon a time they abounded in the royal chases. Grass-snakes, adders, and slowworms may be observed in places suited to their mode of life.

Of birds there is an abundance. The line of the Avon valley is a well recognised route which migrating birds follow. The large artificial reservoirs near Barnt Green lying at the headwaters of its tributary the Arrow afford a resting place for many sorts of water-fowl and in their neighbourhood bird-catchers from Birmingham may generally be seen pursuing their trade on the roadsides unhindered, a public scandal. On summer nights the corncrake, nightjar, and nightingale may be heard. The kingfisher can be seen by most streams and frequents even the canal sides. The presence of the blackcock as far south as the Wyre Forest is worthy of note.

The Severn is famous for its salmon, lampreys, and lamperns. The two latter are considered great delicacies. As already noticed, these fish never run up the Avon. Trout are also caught in some of the rivers. The Teme is the best trout stream in the county, and grayling are also caught in its waters. In all the rivers and canals there is abundance of coarse fish, and Avon eels are said to command the best price in London.

9. Climate and Rainfall.

By climate we mean the average weather experienced in any given region. This, it is evident, is a most important factor in the case of an agricultural county like Worcestershire. The climate of England is every-where changeable, yet the average of the changes is not the same for all parts. The western counties receive more rain and are usually warmer than the eastern ones. It is our present duty to enquire into the cause of such variations and in particular to examine the nature of the climate enjoyed by Worcestershire.

The causes of variation of climate are many and subtle, but chief among them may be mentioned altitude, the nature of the soil and vegetation, proximity to the sea or mountains, and the direction of the prevalent winds.

When we come to consider Worcestershire, we find that the western part is wetter than the eastern, while the north-eastern portion has the coldest and most bracing climate. The former fact is probably accounted for by the proximity of the Bristol Channel and the hill country of Wales and Herefordshire, while it is only natural that the high ground in the north-east should be generally cooler than the low-lying valleys of other parts of the county. The prevalent winds are from south-west and west and come up the Bristol Channel saturated with moisture from the ocean and pass inland along the Severn valley. This is a most important factor to be reckoned with in regard to the rainfall of the county.

Local influences sometimes make a marked difference in the climate of two adjoining places. A hill slope facing south, for example, in that it receives the sun's rays more directly, is warmer and more genial than one facing north. Again, the tendency of a sandy soil is to promote greater extremes of temperature than a heavy clay one.

The influence of altitude is well displayed near Bromsgrove. Here in the valley the blossom opens and the fruit ripens a full fortnight before it does on the higher slopes of the Lickey Hills some two or three miles

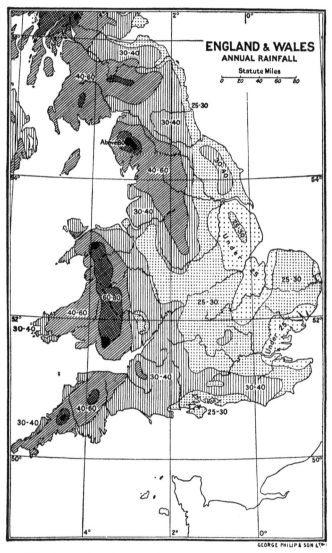

(The figures give the approximate annual rainfall in inches.)

away. This is however the result not only of altitude but of the greater exposure to the wind which the hills suffer. If we compare the Vale of Evesham with these uplands the contrast is even more marked.

One of the great risks in fruit-growing and market-gardening, which are carried out so extensively in the county, is the chance of a late frost or even of snow. If frost should come just when the fruit is "setting" the entire crop may be destroyed. This frequently happens and may spell ruin to the small growers. But we learn here also the importance of local conditions; for an orchard here and there, situated a few feet higher than its neighbours and where the air is dryer and the effects of frost less severe, escapes; and the happy owner makes far more in the bad year than he would have in a good "hit" of fruit. Thus we see that quite small local differences of climate may be of the utmost importance to the inhabitants.

We have now noted some of the influences which the climate exerts, but it may be of interest to compare some of the statistics gathered from meteorological observations over the whole of England with those recorded in Worcestershire. The Meteorological Society collects from numerous stations particulars of the barometric pressure, the temperature of the air, the hours of sunshine, the rainfall, and the direction of the winds. These observations are summarised in the daily papers, so that we can see by means of a chart or map what kind of weather has been experienced over the British Isles. From these observations averages are made out for each year, and we may now compare some of the results deduced from observations in Worcestershire with those collected over the whole of England.

It will be seen from the map (p. 44) that the south-west part of Worcestershire has a greater rainfall than the rest of the county. The average rainfall for Great Britain in 1907 was 33·61 inches while that of Worcestershire was 27·5 inches. This is rather below the average for Worcestershire, which over the period 1870—1899 was 29 inches.

It has already been noticed that the prevalent winds are from the south-west and west. Occasionally they attain extreme violence from these quarters, as in the case of the gale of March 24, 1895, when great damage was done to houses and timber, especially on the exposed summits of the Lickey Hills. The Edgbaston Observatory on this occasion registered a wind-pressure of 37 lb. per square foot, the highest previous record being 27½ lb.

The other meteorological observations are less interesting but it is perhaps worth noting that the mean temperature of the county is 48·6° F. which is decidedly below the average (51·9°) for the western districts of England.

10.　People—Race.　Population.

The earliest inhabitants of Britain of whom we have any written record were the Britons. We have some knowledge of these people from accounts of early geographers who visited this country probably about the fourth century B.C.

The later Roman invasion of A.D. 78 brought with it many of the benefits of civilisation. It did not result in the expulsion of the Britons, for they lived among their conquerors and adopted their civilisation and habits and to some extent their culture. The effect of the Roman

British Camp, Malvern

occupation on the population of Worcestershire was probably unappreciable. For we have only scattered indications of their settlement, and there appears to have been no large Roman town in the county. Probably in its wild and forest-clad condition it was barely worth occupying. In any case we may look upon Worcestershire at the time of the withdrawal of the Romans as sparsely peopled by Britons who had been to some extent influenced by the Roman civilisation.

The huge camps on Malvern, Woodbury, and Bredon, and the ancient roads known as Ridgeways, are the handiwork of these early inhabitants, and give some clue to their mode of life. If we examine the distribution of the remains of this people we are led to regard them as essentially hill folk occupied for the most part as herdsmen and hunters, and paying but little attention to agriculture. Early writers speak vaguely of the inhabitants of this district either as the Silures, Dobuni, Ordovices, or Cornavii but it is doubtful how far any of these names may be relied on. Probably we cannot do more than say that the people formed part of the Kymry who were the forefathers of the present Welsh nation.

In the year 583 the people of Wessex invaded the region we now know as Worcestershire, and defeated the Britons who retreated into the less accessible parts of the forest and to the hill country on the west. The conquerors, who settled down in what is now Gloucestershire, Worcestershire, and Warwickshire, were a tribe known as the Hwiccas. At first they formed part of the large kingdom of Wessex, but later came under the sway of Mercia. Penda, king of the Mercians, waged savage war against the Britons who still remained in the forests of Worcestershire; but on his death the introduction of Christianity as the state religion of Mercia allowed these survivors to remain in comparative peace. Hence it comes about that the county is richer in British place names—for example Malvern (*moel-wern* ?), Pendock,

Cold-comfort (*cœd-cwm-fford*) and the Rhydd (*rhyd*)—than many others, since the Britons were never absolutely exterminated from the region.

The present inhabitants are, however, the descendants of the Angles and Saxons, and not of the Britons. Their very speech betrays them. For the country folk still talk a broad dialect in which many good old Saxon words and endings are interspersed. It is one of the penalties of the advance of education that these archaic forms are rapidly dying out; for it is by the presence of such decaying dialects that we can prove how far the present English language is based on that of our forefathers. If asked whether there is any particular racial type characteristic of the county, we might well answer that, though there is not one conspicuously marked out, yet there is a tendency to darker hair than that prevalent in some of the other midland counties.

In some parts of England we find that the social life and sometimes the nature of the population has been affected by the settlement of skilled foreign artisans. This has been to some slight extent the case in Worcestershire. The most important foreign introduction was that of glass-blowing. This industry was started at Stourbridge about 1600 by immigrants from Hungary and Lorraine and has been in a flourishing state ever since. At one time silk was manufactured at various places in Worcestershire and this trade received considerable impetus from the Revocation of the Edict of Nantes which caused large numbers of skilled workers to come over to this country. In neither of these cases, however, has the nature of the population been affected, though they have made a marked difference in the industries of the county.

Let us now pass on to consider the present population of Worcestershire.   Every ten years a census of the popula-

Glass-houses at Stourbridge

tion of England and Wales is taken, and from the results of the last one, which was held in 1901, we find that there were then 488,338 people in the ancient county of Worcester.   But since part of the ancient county is separated off for administrative purposes, the number of

persons in the administrative county was considerably
less (358,377).

There was a marked increase in the population between
the years 1891 and 1901 and, as would naturally be expected, the greater part of this took place in the manufacturing districts. Thus in the Eastern Parliamentary
division the numbers increased from 59,356 to 96,087,
while in the Northern one, which includes Dudley and
parts of the Black Country, the proportional increase was
only slightly less.

It is surprising to consider the change in numbers
which the population of this county has shown during
the last century. Possibly the census of 1801 was a
phenomenally low one as a result of the drain on the
manhood of the country which the European wars had
brought about, but only 146,441 persons were registered
as being in the county in that year. The growth in
numbers has been fairly constant, for in 1851 we find
a census of 276,927. Still it is extraordinary that the
population should have almost quadrupled itself in the
space of a hundred years. The increase must be ascribed
to the rapid growth of the great manufacturing centre of
the Midlands.

At present Worcestershire has about 605 persons per
square mile. This is well above the average density of
population of England and Wales which works out at
558 people per square mile. For graphic representation
of these comparisons the reader is referred to the diagrams
at the end of the book.

11. Agriculture.—General Considerations. Main Cultivations. Stock.

When we enter upon the study of the agriculture of any particular country we have to keep before our mind's eye certain considerations which are important in determining the nature of the crops produced. Climate ranks first amongst these factors, for it affects not only the plants and animals but even the nature of the soil. We have already seen that the climate of most of Worcestershire is favourable for many crops, though it is too rigorous for a few of the more delicate ones. The sufficiency of rain which it receives makes it an excellent district for growing hay and clover.

Next in importance is the nature of the soil. This varies as a rule with the underlying geological formation from which it is derived. The Teme valley with its light loamy soil, which originates from the Old Red Sandstone Formation of that district, is especially adapted for hop-growing, whereas the Lias Clay gives the heavier soil of the Vale of Evesham, which is well suited for orchards. It is of interest to note that in many cases the fertility of the soil depends on the admixture of material from two distinct geological formations by means of streams and rainwash.

The demand and the price obtainable also help to determine the nature of the crop to be raised. We can find an excellent example of this in the history of the agriculture of Worcestershire. With the high price of wheat which prevailed at the end of the eighteenth century, the cider orchards of the Vale of Evesham were cut down and replaced by corn. At the present day corn is no longer grown profitably on a large scale in England, but the expansion of Birmingham into a great commercial centre has created a demand for fruit and vegetables which formerly did not exist, and now the Vale is one vast orchard and market garden. Again in the last year or two there has been an ever-increasing demand for *early* fruit and vegetables, and it is the same Evesham which has introduced the "intensive" form of cultivation that has long been practised in France. On this system fruit and vegetables are induced by the liberal use of glass protection and manure to grow at the greatest possible speed. By this means and judicious planting, crops of early cabbage, lettuce, asparagus, etc. are put on the market two or three weeks before the ordinary time and much higher prices can be realised.

We may here note the influence exerted on the nature of the agriculture of a region by the means of transport available. In the case of Worcestershire the railways are at present sufficiently ramified to allow even quickly perishable produce from most districts to reach the market at Birmingham, London, South Wales, Lancashire, and Scotland. A very different state of things prevailed about the year 1800, when communications were very poor and the material was mostly carried by pack-horse. It is however curious to note that the people of those days made great improvements in the roads, as

Intensive Cultivation near Evesham

they found it profitable to abandon the pack-horse and adopt carts. At the same date, we learn, the Trent and Severn Canal carried quantities of certain kinds of fruit; as much as 7000 tons going over it in a year when there was a "hit of fruit." Still it is evident that perishable market garden produce such as strawberries, lettuces, and asparagus were then of no use except for local consumption.

When we turn to a more particular examination of the agriculture of Worcestershire at the present day, we derive considerable help from the Returns of the Board of Agriculture in which the annual agricultural statistics are collected and tabulated. The usual method of estimating the crops is by the acreage they occupy. This has a disadvantage in that it conveys no idea of their actual or relative values.

Most of Worcestershire is under cultivation. When we consider that large areas were, until comparatively recently, forest land and that other parts lie too high to be very fertile, it is surprising to find that there are only a few acres of waste land and 21,768 acres of woods. If one looks from a commanding point, one would say that Worcestershire is a well-wooded country, but this actual acreage of woodland is small and the effect is produced by the abundance of trees in all the hedgerows.

The Board of Agriculture classes wheat, barley, oats, rye, beans, and peas together as "Corn Crops." In 1907 there were 71,338 acres under these crops in Worcestershire. This was about one-fifth of the total acreage under crops and grass, which amounted to 399,582 acres. Peas and beans form a prominent part of the corn crops of the county.

The "Green Crops" include potatoes, turnips, swedes, mangolds, cabbage, rape, and vetches or tares. These occupied about 27,000 acres, potatoes being the most important item. Only a small portion of the county is given over to "Rotation Grasses." These are clover, sainfoin, and grasses, sown in rotation with corn crops.

A large part of the county, amounting to more than half of the total area under grass and crops, is in permanent pasture. About two-fifths of this is mown for hay while on the rest cattle and sheep are grazed.

Worcestershire is one of the few hop-growing counties. In 1907 the hop gardens, which lie mostly in the Teme valley or near Worcester, covered 3622 acres. This allows Worcestershire to rank sixth among the hop counties. The picking of the hops is mostly done by people from the Black Country, for many of whom it is a great holiday. Others spend the summer months pea-picking, hay-making, and fruit-picking, and then move into the hop-fields as autumn arrives. In some few places the inhabitants do the hop-picking and are known as home-pickers.

Turning now to more specialised crops, we must note at once 21,385 acres of orchards. This combines with the 4790 acres of small fruit to constitute Worcestershire's claim to rank as one of the Gardens of England. A large proportion of the apples and pears grown in this county is employed to make cider and perry. These drinks were formerly made with great care, and before wine was so

Hop-Garden in the Teme Valley

plentiful the better varieties formed good substitutes for it ; but at the present day far less is produced and this is more often manufactured on a large scale, whereas formerly every farmer made his own.

Even so far back as the reign of Henry III, Robert of Gloucester sang the praise of the "frut of Wircestre" and three pears appear natural in the arms of the capital of a county so famous for its fruit as Worcestershire.

A Hop Room

At the present day there is an ever-growing tendency to cultivate the finer varieties of all fruits.

Pershore is especially noted for its plums and damsons, although the whole of the Vale of Evesham abounds in them. Here the owners combine market-gardening with fruit-growing. For the soil is so fertile that it is possible to have several crops on the land at

once. Thus lettuces and wallflowers grow between
tomatoes or gooseberries, which in turn fill in the spaces
between rows of plum trees.

Asparagus and other vegetables are also largely grown
in the Avon valley and command good prices in London
and Birmingham, in fact asparagus is sent all over England
from Evesham. It has been found that the slopes of the
Lickey Hills are especially well suited for strawberry cultivation, and they provide a large part of the supply for
Birmingham. In the whole county there are some 1200
acres devoted to strawberries.

The Returns of the Board of Agriculture also tell us
that there were 21,909 horses and 73,442 head of cattle
in Worcestershire in 1907. Sheep numbered 160,516
while 45,536 pigs complete the list. Of recent years
large poultry farms have been started near Bromsgrove
and Redditch.

12. Industries and Manufactures. (*a*) Those dependent on Natural Resources.

When we come to consider the industries and manufactures of any country, we find that there are two chief
factors at work in determining their nature. In the first
place they must depend on the natural resources of the
country, and in the second on the adaptability and in-
clinations of the inhabitants.

Now Worcestershire has so far been well favoured in

both these respects. Its chief resources are salt, coal, ironstone, limestone, sandstone, and clays, especially the famous fireclays of Stourbridge. In former times the natural wealth of forest was another aid to prosperity ; for before the use of coal was introduced by Dud Dudley, the forests provided fuel for the salt pans and blast furnaces.

It is only natural that the geographical distribution of the resources should determine the position of the industrial centres. In Worcestershire these are chiefly found in the north-western district on the border of the South Staffordshire coal-field ; but Droitwich forms the best possible example of a localised industry dependent on the position of the natural supply.

In other cases the presence of means of communication or of water-power may determine the position of an industry. Instances of this can be multiplied from this county. Thus Bewdley in the days of its cloth manufacture lay on one of the chief trade-routes to Wales, and so became a centre to which Welsh wool was brought. Again, in the " Canal Era " at the end of the eighteenth century, Stourport sprang into commercial activity on account of its position at the point where the Trent and Severn Canal joined the latter river. Nevertheless in Worcestershire there are several instances of specialised industries confined to particular localities of whose existence it is not always easy to see any explanation. A case in point is the porcelain industry at Worcester ; for the town produces none of the ingredients necessary for a manufacture which has flourished there for 150 years.

Salt-making is so essentially a Worcestershire industry that it will be necessary to consider its history and methods in a separate chapter.

The smelting of iron has been in operation round Dudley for centuries, and it was here that coal was first used for this purpose by Dud Dudley about the year 1621. In times past the iron ore was obtained from the local Coal Measures but at the present day it is chiefly brought from other districts, notably Northamptonshire, with the result that the industry is on the whole a declining one as an outcome of the exhaustion of the local supplies. But the prosperity which has arisen from this industry has made the Dudley area a centre for related trades. Chain-making flourishes chiefly at Cradley, while other manu-factures are railway bridges, constructional iron and steel work, railway carriages and wagons, anvils, edge tools, nuts and bolts, iron pipes, safes, fenders, and fire irons. The manufacture of hand-made nails, which was formerly carried on on a considerable scale, especially at Bromsgrove and Halesowen, has been almost displaced by the intro-duction of the machine-made article. A curiously localised trade, the manufacture of scythes, has been carried on for more than a century at Belbroughton.

Brass foundries and metal-rolling works are found in the Dudley district and near Birmingham. The ex-tension of the latter town into Worcestershire, especially near Selly Oak, might allow us to mention other hardware trades which are carried on here. Suffice it to say that they include bicycle, motor-car, gun, cartridge and ammu-nition, and enamelled-iron manufactures.

Stourbridge and Oldbury are famous for their glass. The industry was founded at the former place by glass-blowers who had immigrated from Lorraine. Their choice of Stourbridge was due to the exceptional quality of the fire-clay which is found there and which is used to make the crucibles in which the glass is melted. The photograph on p. 50 illustrates the conical shape of the glass-houses. Plate and window glass is chiefly made at Oldbury and here too a very specialised and world-famed branch of the industry is carried on, namely the casting, grinding, and polishing of the huge lenses for light-houses.

An allied trade connected with glass-making is the manufacture of glazed bricks and porcelain baths. This is also in operation in this district, as also are important chemical works for the production of alkali and phosphorus.

The fire-clay is used for making fire-bricks with which blast and other furnaces are lined. The presence of this fire-clay in the neighbourhood of an iron smelting district cannot have been without advantage to the latter industry, which thus had the necessaries for smelting—the ironstone, the coal, the Dudley limestone and the fire-bricks—all at hand.

We have already made mention of the cider and perry manufacture to which the orchards contribute and now may note that there are breweries in most of the large towns. Boots and shoes are made at Bromsgrove and Worcester, and tiles and bricks at various places.

The manufacture of "Worcestershire sauce" is an

industry dependent on the malt vinegar trade. This latter is one of the most important ones pursued at Worcester, the average yearly output being no less than 2,000,000 gallons.

13.   Industries   and   Manufactures. (*b*)   Those   not   dependent   on Natural Resources.

Let us now revert to those isolated industries mentioned above which are found at various places where there appears to be no very clear reason for the choice of locality.

Worcestershire may be said to have an almost world-wide monopoly in the production and sale of needles and fish-hooks. The seat of this industry is the town and neighbourhood of Redditch, and practically all English-made needles and fish-hooks are manufactured here. There were in the eighteenth century needle-makers in Birmingham but about 1820—1830 they all migrated to Redditch, though the reason for their choice is not obvious except in so far as the river Arrow provides some water-power. At the present day there are about 2000 hands employed in this trade.

The chief manufacture carried on at Kidderminster is that of carpet-making, which employs some 5400 hands. Woollen goods have been made here for centuries, and we can trace the rise and fall of several successive types of manufacture, showing that the people of Kidderminster

have adapted themselves to new demands, whereas other
equally ancient woollen centres such as Bromsgrove and
Bewdley have now almost completely lost the trade. At
Kidderminster, the first industry of which we have any
record was that of wool and cloth manufacture. As the
trade declined, a mixed stuff known as Linsey-woolsey
was made, which in turn gave place to silk. In 1755
there were 1700 silk-looms in Kidderminster, but the

Worcester China

number had shrunk to 340 in 1831. About this date
the manufacture of carpets began and has flourished until
the present day. "Axminster," "Wilton," and "Brussels"
are the chief varieties manufactured. The only reason
that can be assigned to the localisation of these trades
at Kidderminster is its proximity to Bewdley, formerly
an important wool market to which Welsh wool was
brought.

In Worcester, the chief industries are glove and porcelain making. The origin of the latter in this town is of interest. As its woollen industry declined, certain citizens, especially one Dr Wall, about the year 1750 sought for some new one to take its place, and chose that of porcelain. Fine china only was made at the works then established, which after the visit of

Bournville

George III became known as the Royal China Pottery. Other potteries which were then or later in existence have now for the most part been amalgamated and the Royal Porcelain Works still carry on an extensive business in the finer varieties of porcelain.

The manufacture of cocoa and chocolate which, in our county, is carried on only at Bournville near Birmingham must also be mentioned. This is of especial interest on account of the social schemes which have originated in connection with it and which have led to the erection of a model village for the workpeople.

We will conclude this chapter with a reference to one quite small industry which on account of its great antiquity must not be omitted.   This is the manufacture of parchment which, started at Bengeworth near Evesham for the monks of the latter place, was carried on there until about 1830.

14. Salt=making at Droitwich.

One of the most ancient industries of Worcestershire is that of salt-making, which has been in existence from time immemorial at Droitwich.   It is exceedingly probable that the salt springs were known to the Romans, for there was a small settlement here at that date.   There is also a road leading from Droitwich in a north-easterly direction, known as the Upper Saltway, which is thought to be Roman, though it may quite well be that it follows the line of a more ancient British trackway.   Other saltways are also recognisable which are probably in some instances of British origin, so that it is more than likely that the industry was in existence at Droitwich even before the Roman occupation.

The right to make salt was highly prized by the early English inhabitants and some authorities believe that the word "*wich*" means a salt-pit or salt spring, and even connect the tribal name of Hwiccas with it.   In the

Domesday Survey we gather abundant proof of its value, for not only landowners and towns in Worcestershire but also others at a distance possessed this right and the king claimed monies for it from the burgesses of Droitwich. These continued to possess a monopoly of the brine from the salt springs, which were until 1725 the sole source of the supply.   The amount of brine was very limited and

Dodderhill Church and Salt Works, Droitwich

each man's portion appears to have been given out to him by officials.   This scarcity and, more especially, the monopoly enjoyed were the causes of the high value of the salt.   At the same time the crown claimed a heavy tax on it which was only removed in 1825.   Nash, the historian of Worcestershire, states that the year's output for 1771–72 was 604,579 bushels and that the duty paid into the salt

office at the rate of 3s. 4d. per bushel amounted to
£61,457. He estimates also that with this grievous
duty "the labourer who salts his pig, pays above 15s."—
for what would be worth 5d. were the tax removed.

At the present day the brine is obtained exclusively
from deep borings at Droitwich and Stoke Prior, and there
is no monopoly. Salt was first found at the latter place
in 1828. The supply from the deep wells is abundant
and naturally is dealt with on a much larger scale than
formerly. In 1906 there were 168,486 tons of salt
produced, the value of which was £75,812. At the
time of the Conquest, leaden pans were used for evaporating the brine; the forests provided the fuel, and during
the middle ages the industry threatened to destroy the
neighbouring forest of Feckenham. Thus Camden writes
of it :—"What a prodigious quantity of wood these saltworks consume, though men be silent, yet Feckenham
forest, once very thick with trees, and the neighbouring
woods, will by their thinness declare daily more and
more." Later, iron pans and coal were used.

The present method of treatment is as follows :—the
brine is pumped from the wells into an open brick
reservoir from which it is distributed by pipes to the
evaporating pans. These are of two kinds ; the smaller
measure 40 × 22 ft., and in these the evaporation is carried
on rapidly, the result being the formation of a finer grained
salt than that which is obtained by slow evaporation in
the larger salt pans measuring 80 × 22 ft. The latter
variety is known as "*broad salt*." During the process
a crust forms, then sinks and is raked to the side and

placed in wicker "tubs" or moulds while hot and wet.
It drains in these and sets hard enough in 20 minutes
to allow the "*bar*" to be completely dried elsewhere.
These bars are the almost sugar-loaf shaped ones in which
salt is usually sold.

It may be profitable to glance at some of the effects
which have been produced by the presence of this natural
resource hidden deep down in the rocks below Droitwich,
and stored there by the drying up of some inland salt-
lake in Triassic times, long before man appeared on the
earth.   In the first place we have already mentioned
its effect on the neighbouring forests, which it has consumed.   It has kept men employed for centuries in
extracting it ; it has caused roads to be built and canals
to be cut ; it has made Droitwich a spa of considerable
importance where many rheumatic patients derive great
and often lasting relief from treatment by brine baths ;
lastly it has done all this without that disfigurement of
the landscape which so often results from the development
of such products as coal or iron.

15. Mines, Minerals, and Water Supply.

The last chapter has dealt with the salt-making
industry at Droitwich, which might very well have been
considered in the present chapter ; for, although the salt
is not mined in the ordinary way, yet it must be regarded
as one of the most important mineral resources of the
county.

It is only in the northern extremity that coal has

been found in workable quantities. At Dudley and Cradley the famous 30 ft. seam of the South Staffordshire coalfield was at first mined at the surface. It has however now been practically worked out. South of Cradley the coal seams appear to thin out along a line drawn between Birmingham and Stourbridge. The Bewdley or Wyre Forest coalfield also consists for the most part of unproductive measures and coal is now only worked at four small pits in this district.

There has been much talk of late years of hidden coalfields. By this is meant coalfields covered by strata newer than the Coal Measures below which they have to be worked. There is very little doubt that Coal Measure strata occur below newer rocks between the South Staffordshire and Bewdley coalfields. Since however the southern part of the former and much of the latter are unproductive of workable seams it is probable that even if we sank a pit through the Triassic rocks and found Coal Measure strata at a workable depth, they too would be unproductive. This is unfortunate for the county; for, farther to the north, thick coal has already been found below the Triassic covering, several miles to the west of the present edge of the South Staffordshire coalfield; and as the supply from other places decreases, these hidden coalfields will become gradually more valuable and profitable to work.

There are, however, other useful products from the Coal Measure strata of the northern part of Worcestershire. Of these the ironstone has been practically worked out and imported ores are now chiefly smelted.

Stourbridge fire-clay is also a Carboniferous deposit. It is used to make crucibles for glass-making and fire-bricks for lining furnaces, and is said to derive its reverberatory properties from the absence of alkalies and from a high percentage of silica. It is generally supposed that fire-clays are the remains of old land-surfaces on which the Carboniferous plants grew, and that it was these plants which abstracted the alkalies for their tissues, thus adapting the clays to their present uses.

As has been mentioned above, there were at one time in the Dudley district the coal as fuel, the ironstone nodules as ore, the fire-clay for lining the furnaces, and the limestone to form the flux—all the essentials for iron smelting. (The Silurian limestone at Dudley lies just outside the county boundary but is of sufficient importance to the iron trade to be mentioned here.) Worcestershire was, however, an iron-smelting county long before coal was used as fuel. In those early days the forests here, as in the Weald of Kent, provided the charcoal with which the iron was smelted. The furnaces were then distributed along the Severn valley and near Dudley.

In the south of the county the Lias limestone was formerly extensively quarried for lime-burning and for the blue flagstones with which so many of the older farms and dairies were paved. The use of these "flags" has now been given up, as they have not proved themselves sufficiently durable and at the same time have the disadvantage of being always damp during wet weather. The Lias quarrying industry is now nearly extinct.

Worcestershire is well supplied with building materials. The forests are now gone, but the abundance of half-timbered houses is testimony that at one time the material for wooden buildings was not wanting; these beautiful half-timbered houses date from a time when bricks were not so cheap as they are now. The red bricks

Mere Hall, near Droitwich
(*Ancient half-timbered house*)

and tiles which give the Worcestershire houses their fine warm colours are made either from the Triassic marls as at Yardley and Alvechurch, or from the Glacial clays as at Blackwell. In mediaeval times Malvern and Droitwich were famous for their ornamental floor tiles. Examples of these exist in the priory church at Malvern and show the fine quality both of design and colouring which won their renown.

The Worcestershire churches are mostly built of grey or red Triassic sandstone. At the present day this is only quarried at a few places.

Road metal is obtained from the Pre-Cambrian rocks of Malvern, the Cambrian quartzite of the Lickey Hills, from the Dolerite of Rowley Regis, while some gravel is derived from Glacial beds.

The effect of mining industries is usually to alter and disfigure the landscape. Sometimes it is gaunt tip-heaps which mar it, sometimes the vegetation is destroyed for miles round, and, sometimes, subsidences take place at the surface where the ground has fallen in as the mineral wealth has been taken out. This last has occurred to a considerable extent at Droitwich, while the Dudley-Cradley district has been entirely disfigured by the spoil-heaps or "pit banks" of the coal mines and blast furnaces. Perhaps, in a few generations, afforestation will have clothed these unsightly heaps with vegetation and brought back some of its former beauty to that district. Many of the older tip-heaps are composed of slag from the blast furnaces. Happily the tendency of the present age is to find some use for what was formerly considered a waste product, and slags, for example, are now employed for various purposes, those which contain phosphates, after being finely ground, being used for manure, while others are broken up for railway ballast and road metal. To a certain extent this diminishes the size and extent of the present waste-heaps from blast furnaces.

An account of the natural resources of a county would not be complete without mention of the water-supply; for the nature of the water necessarily affects the lives of the inhabitants.   In those parts of Worcestershire which are underlain by Triassic sandstones, very good drinking water with a slight permanent hardness is obtained from these water-bearing strata.   In other parts springs and shallow wells are generally employed, while the city of Worcester derives its supply from the Severn.

16.   History of Worcestershire.

Until the seventeenth century the geographical position of Worcestershire on the west side of Mid-England, virtually forming as it did a border-county against Wales, constituted the chief determining factor in its history.   Its position led it to being the scene of continual struggles with the Welsh in the early days when the line of the Severn with its swamps and tidal stream formed the real border of Wales on the east. The Severn was undoubtedly the most important strategic feature, and whoever commanded it could check the piratical raids of the Irish and Danes and keep the ever restless Welsh in their own country.

At the time of the Roman invasion under Claudius in A.D. 43, the Britons occupied the forests of this region, and the numerous camps and earthworks probably indicate that they disputed possession with the Romans.   But in

any case the latter were victorious. Their border against
the Britons, however, lay farther to the west in Herefordshire and we can find no traces of their military occupation of Worcestershire.

In the year 583 the West Saxons under Ceawlin and
his brother Cutha appear to have driven the Britons
out of the lower Severn valley. We know but little of
the gradual settlement by which the English peopled
Worcestershire but it is reasonable to argue from the
abundance of Welsh place names, especially in the south-
west of the county, that the Britons lingered on here
later than in most parts of England. The tribe of the
Hwiccas peopled Gloucestershire, Worcestershire, and
Warwickshire, owing fealty at first to the West Saxons
and later to the Mercians.

St Augustine travelled through this part of England
and somewhere hereabouts may have held the famous
council with the Welsh bishops. But the real adoption of
Christianity with its wide-reaching civilisation occurred
after the death of Penda, king of the Mercians, when his
son Wulfere, who was a Christian, became king. St Chad
was ordained first bishop of Lichfield, but that see proving
too large, the diocese of Worcester was separated from it
in 680. The first bishops of Worcester, or bishops of the
Hwiccas as they were called, came from Whitby, which
in turn had been founded by missionaries from Lindisfarne,
so that this part of England cannot be said to owe its
Christianity to St Augustine.

An appreciation of the power of the Church in
Worcestershire is a most important key to the right

The Priory Church, Little Malvern

understanding of the history of the county; for from the earliest times the Church has played a most prominent part—a fact of which it is important to grasp the significance at once. Until the Dissolution, the monasteries at Worcester, Evesham, Pershore, and Malvern, besides the abbey of Westminster and a host of smaller religious houses, were the chief landowners in the county.

The establishment of the Benedictine rule in the larger monasteries, brought about by St Oswald, bishop of Worcester, about the year 977, was the true beginning of the present agricultural prosperity of the county. For this Order seems always to have proved successful in converting barren lands into orchards, cornfields, and vineyards.

The history of Worcestershire in these early days is really comprised in that of the town of Worcester, which was repeatedly destroyed by fire at the hands of the Danish pirates. From the reign of Ethelred until the time of Henry II there was a mint at Worcester.

It was fortunate for the county and for England at the time of the Norman Conquest that they possessed such a man as St Wulstan "the Englishman" as bishop of Worcester. After the battle of Hastings when further resistance was futile, he gave his allegiance to the Conqueror. His loyalty saved most of the church lands from confiscation, despite the rapacity of the sheriff Urso d'Abitot, and it even led him to co-operate with the sheriff in suppressing the revolt of the barons in 1074.

In 1088 Worcestershire was once more the scene of conflict between the barons of the Welsh Marches and

Pershore Abbey

the loyal forces of bishop and sheriff, and again the "loyal city" prevented the success of the rising.

Worcestershire was the scene of much trouble in the civil wars of Stephen's reign. The town remained loyal to the king and was repeatedly taken and retaken. One of Stephen's great advisers was Waleran, Earl of Worcester.

The building of the castle at Worcester marks the Norman occupation, but the decline of the lay power is seen in the fact that before 1220 the castle at this place was partly in the hands of the church: thus the bishops in these early times were temporal as well as ecclesiastical lords.

King John appears to have been especially attached to Worcester and often came here to hunt in the royal chases, which in the middle ages were governed by a special code of laws which Hallam says constituted "an oasis of despotism in the midst of common law."

John in his will specified that he should be buried in the cathedral at Worcester, and his tomb may still be seen there between the shrines of St Wulstan and St Oswald.

Some idea of the importance of the bishops may be gleaned from the fact that the history of the county between 1237 and 1266 is centred round Bishop Walter de Cantilupe, who was a strenuous supporter of the barons' cause in the war against Henry III. It was probably due to his influence that in 1265 Simon de Montfort managed to cross the Severn at Kempsey when all the other fords were in the hands of the Royalists. It was on the next day at Evesham that the bloody battle was fought in which Earl Simon was slain and the baronial power shattered.

From this time until the beginning of Henry VII's reign, Worcester served as a base for operations against the Welsh, who were a continual source of unrest to the West of England. The institution of the Court of the Welsh Marches dates from the time of Edward IV. Ludlow and Bewdley were both places at which it sat and its jurisdiction embraced Worcestershire. In later times the conduct of this court was very tyrannous, and fell as a severe burden on the county.

The dissolution of the monasteries by Henry VIII did not bring about so great a change in Worcestershire as elsewhere. For, although the numerous religious houses were all closed, the bishop of Worcester kept his lands, and the possessions of the abbot and convent of Westminster in the county eventually passed into the hands of the dean and chapter of that Abbey. Nevertheless the county was so rich in such houses that the change appears great; for we find Pershore and Evesham and a host of minor establishments passing into lay hands. The result of this was that the magnificent churches and monastic buildings were either completely demolished or some small fragment alone was saved by the public spirit of the inhabitants.

The advent of the Stuarts brought trouble to Worcestershire. The oppression of the Catholics under James I led, among other things, to the Gunpowder Plot, in which many Worcestershire families were implicated,

King Charles's House, Worcester

often with disastrous results. The two Wintours were among the chief conspirators, while the Lytteltons, Stephen of Holbeach, and Humphrey of Hagley joined them later. All paid the penalty with the loss of their lives and property.

Throughout the Civil Wars we find Worcester a centre of military operations. In 1642, the first year of the war, the town was in the hands of the Royalists, who opened the campaign by the brilliant victory of Prince Rupert and his cavalry at Powick. But it was captured the same year by the Parliamentarians, only to be shortly re-captured by the Royalists, who kept possession until the end of the war. It only capitulated after a two months' siege when Charles's cause was hopeless. The Royalist garrisons of Dudley, Hartlebury, Madresfield, and Strensham held out almost as long.

During the war the county suffered fearfully, contributions being levied by both sides. Continual raiding also took place and it is doubtful which party was the more unpopular.

In the second Civil War, the " faithful city " was the scene of the last struggle of Charles II to recover his crown. In 1651 he marched south with a Scotch army to Worcester, where on September 3rd he was defeated by Cromwell and compelled to fly in disguise to France. The fight began at Powick on the west side of the Severn but later centred round Fort Royal and the Sidbury gate. That evening Cromwell wrote to Parliament, describing the victory as a " crowning mercy."

From this time onwards Worcestershire has been free

from the alarms of war and has expanded its industries, both agricultural and commercial, with considerable success, as is evident from its present state of prosperity.

17.   Antiquities—Prehistoric.   Roman. Saxon.

In this chapter it is proposed to carry the mind back to the dim past and consider what is known of this district and its inhabitants before the dawn of written history.   For here, as is everywhere the case, the early peoples have left traces of themselves in the shape of antiquities, and since no very early written records of Worcestershire exist we have to rely on such archaeological evidence for information about the peoples who inhabited this district at a time when, in other regions, written histories were being compiled.   Thus, although we know, for example, that the Romans conquered this district, yet there is no written record of it, or of any place in it, before the sixth century A.D.

Antiquaries have divided the earliest portions of our country's history into the Stone Age, the Bronze Age, and the Iron Age.   After this last period we generally speak of Roman antiquities and Saxon antiquities, which correspond with the history of our land from 55 B.C. to 1066 A.D.   Antiquities representing all these divisions of time have been found in Worcestershire but the county is remarkable for the paucity of such remains.

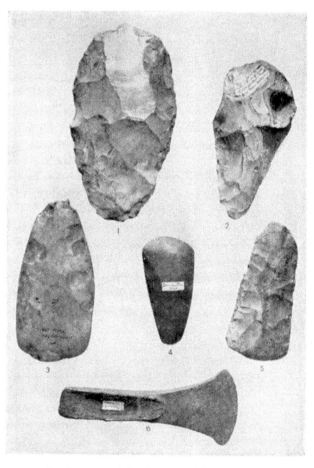

Implements of the Stone and Bronze Ages

1, 2 *Palaeolithic,* 3—5 *Neolithic,* 6 *Bronze-age Palstave*

There appear to have been several successive waves of invasion and immigration in England, in which the better armed invaders displaced the former inhabitants. In the first of these the original inhabitants, who still employed only stone implements, were driven out or destroyed by the men with bronze weapons. These invaders were the Goidels, who in turn fell victims to the iron swords of the Brythons. The latter were the people whom we generally speak of as ancient Britons and were the inhabitants who were conquered by the Romans.

Turning now to Worcestershire in particular, we have already remarked on the singular absence of prehistoric antiquities. This is probably the result of the sparseness of the population in the dense forests which then covered the district. Early man appears to have dwelt chiefly in the southern parts of the county, where the Avon valley formed the route along which successive invasions proceeded. The line of the Severn naturally provided a rough boundary between the invaders and the older inhabitants whom they had succeeded in driving to the west. Later, however, the Saxons and Danes made their way into the country chiefly by working up the rivers in their boats, so that the boundaries came to lie rather along the watersheds than along the river courses. Consequently the line of hills on the western side of Worcestershire acquired as a boundary a strategic importance which the extensive series of earthworks on its summits would have led us to expect.

The weapons which were made by Palaeolithic man— man, that is, of the Early Stone Age—were exceedingly rude, being in the main roughly chipped from flint and not ground or polished, while the succeeding race of Neolithic men shaped the stones better, and often ground and polished them. So far no Palaeolithic implements have been found in Worcestershire, but the occurrence of Neolithic implements, for the most part on the hills, though occasionally in river silts, seems to point to the men of that period having been herdsmen who lived on the uplands, which were probably the only parts of the land not densely clothed with forests.

But few traces of the Bronze Age people have been found in the county, and none in the northern and eastern parts. In this district it is the large earthworks which cap the western hills and other eminences and which are distinguished by their round or oval form, that supply the most remarkable evidence of the people of this period. Their roads also may in many cases still be traced. One set is found to connect the large camps, while others appear to be related to the Droitwich salt industry and are known as the "Saltways." The British roads are chiefly remarkable for the preference they show for the hill tops, and from this fact they derive their name of "Ridgeways."

When we come to consider the few Roman remains which are known in Worcestershire we must remember that the county then formed no sort of a political unit, and in order to understand the conditions which existed at the time we must regard the district merely as a part of the Roman province. The term Romano-British is really better than Roman, for the chief effect of the occupation of the country was the civilising of the inhabitants, who in the course of time were admitted into the administration and had to serve in the army.

Now although the conquest of the southern part of England was relatively speedy, yet the hill-country of

A Severn Coracle

Wales and northern England was long in an unconquered state, which necessitated a military occupation. Worcestershire formed part of the former area and we have no evidence of large military camps or stations in the county.

Worcester and Droitwich were both small towns, but the rest of the occupation was in the form of scattered estates or villas very often owned by Romanised Britons. These villas are rare in the county, but there are various places where Roman pottery and coins provide indications of settlements.

The strongest proof of the thoroughness of the Roman occupation of England is afforded by the magnificent roads which they constructed for military purposes. As a rule these run with extraordinary straightness from place to place. We have in Worcestershire several fairly well marked local roads and parts of two big trunk roads, the Icknield Street and the Fosse Way. The latter only touches detached parts of Worcestershire but the former runs through the eastern portion of the county.

Saxon antiquities are almost unknown from the county though a few have been unearthed in the Evesham valley, among them being a peculiar censer assigned to this period.

The Severn coracle, a primitive boat of interlaced framework coated with skins or tarpaulin, which can be easily carried on the back, is still used for fishing. It is a type, without doubt, which dates back to a very remote age.

18. Architecture. (*a*) Ecclesiastical— Cathedral, Churches.

When we consider the architecture displayed in the three main types of buildings in Worcestershire, viz. (*a*) Ecclesiastical or buildings relating to the Church, (*b*) Military or Castles, (*c*) Domestic or houses and cottages, we find that the abundant supply of good building-stone has been taken advantage of by the wealthier classes, while for the smaller domestic houses it has not been much employed. The latter are for the most part red brick constructions, though in the older ones brick has been combined with wood to make the "half-timbered" houses

Old Cottages at Cropthorne

and cottages so characteristic of these west midland counties. So we see here as elsewhere that the type of accessible material is not without its influence on the architecture of the country.

We have pointed out how influential the Church was in the early history of Worcestershire. Accordingly it is only natural to find that the county was rich in churches and abbeys. Those of Worcester, Pershore, Evesham, and Malvern were the largest, and their remains show clearly the magnificence to which they attained.

The churches of Worcestershire are of various styles and of different ages, but before examining them more closely we must glance at the characteristics of the buildings of the various periods.

For all practical purposes the orders of architecture with which we have to deal are the following, which, with the exception of Perpendicular, merged into each other by gradual transition, without sudden break, so that it is difficult to say where one ended and another began.

Pre-Norman or—as it is usually, though with no great certainty termed—Saxon building in England was the work of early craftsmen with an imperfect knowledge of stone construction, who commonly used rough rubble walls, no buttresses, small semi-circular or triangular arches, and square towers with what is termed "long-and-short work" at the quoins or corners. It survives almost solely in portions of small churches.

The Norman Conquest started a widespread building of massive churches and castles in the continental style called Romanesque, which in England has got the name of "Norman." They had walls of great thickness, semi-circular vaults, round-headed doors and windows, and massive square towers.

From 1150 to 1200 the style of building became lighter, the arches pointed, and there was perfected the science of vaulting, by which the weight is brought upon piers and buttresses. This method of building, the "Gothic," originated from the endeavour to cover the widest and loftiest areas with the greatest economy of stone. The first English Gothic, called "Early English," from about 1180 to 1250, is characterised by slender piers (commonly of marble), lofty pointed vaults, and long, narrow, lancet-headed windows. After 1250 the windows became broader, divided up, and ornamented by patterns of tracery, while in the vault the ribs were multiplied. The greatest elegance of English Gothic was reached from 1260 to 1290, at which date English architectural sculpture was at its highest, and art in coloured glass-making and general craftsmanship at its zenith.

After 1300 the structure of stone buildings began to be overlaid with ornament, the window tracery and vault ribs were of intricate patterns, the pinnacles and spires loaded with crocket and ornament. This later style is known as "Decorated," and came to an end with the Black Death, which stopped all building for a time.

With the changed conditions of life the type of building changed. With curious uniformity and quickness the style called "Perpendicular"—which is unknown abroad—developed after 1360 in all parts of England and lasted with scarcely any change up to 1520. As its name implies, it is characterised by the perpendicular arrangement of the tracery and panels on walls and in windows, and it is also distinguished by the flattened arches and the square arrangement of the mouldings over them, by the elaborate vault-traceries (especially fan-vaulting), and by the use of flat roofs and towers without spires.

The mediaeval styles in England ended with the dissolution of the monasteries (1530–1540), for the Reformation checked the building of churches. There succeeded the building of manor-houses, in which the style called "Tudor" arose—distinguished by flat-headed windows, level ceilings, and panelled rooms. The ornaments of classic style were introduced under the influences

Worcester Cathedral—Crypt

of Renaissance sculpture and characterise the "Jacobean" style, so called after James I. About this time the professional architect arose. Hitherto, building had been entirely in the hands of the builder and the craftsman.

In the case of Worcester Cathedral the construction covered all these periods to the Perpendicular, but unfortunately every trace of the Saxon Church has been

Worcester Cathedral—Nave

destroyed. But it may be of interest to glance at the Cathedral as illustrative of the various styles.

The crypt, one of the earliest in England, which was built by St Wulstan, and some of the arches in the transept, which may be easily distinguished by their being round and ornamented with simple geometrical patterns, give a good idea of the scope of the Norman type.

Worcester Cathedral — Choir

The choir belongs to the Early English style and was commenced in 1224. It is a very good example of the beauty attained by the architects of this period. Most of the nave belongs to the succeeding or Decorated Period.

Prince Arthur was buried in Worcester Cathedral, and above his tomb a wonderful chantry chapel was raised by his father, King Henry VII. This monument belongs to the best style of the Perpendicular Period. Its graceful slender shafts and its airy window spaces are at once attractive and distinct from earlier styles (p. 127).

Ruskin has called attention to the intimate connection that existed between architecture and sculpture throughout the Gothic period of architecture, and Worcester is fortunate in possessing good proof of this. For during the Commonwealth the rich carvings which adorn the choir and eastern transepts were all plastered up and so escaped the destruction which usually overtook such work throughout England. They have now been laid bare, and we can see that the sculpture indicates workmanship of no mean excellence. Some of the groups are grotesque, but they all live. The most beautiful of them, a small crucifixion, is one of the most exquisite examples of such work in the kingdom.

Having thus gathered a rough idea of the nature of the architecture of these main styles, we are more fitted to examine the churches of Worcestershire, few of which were built entirely at one period.

A striking feature of Worcestershire churches is the warm tones which their exteriors possess. They are mostly built of sandstone which, be it red or grey, is soon lichen-covered and moss-grown. Some districts in England show a general uniformity of type in their church architecture, but this is not well marked in Worcestershire.

There are no examples of Saxon churches in the county, though parts of those at Rous Lench and

Ribbesford are probably of this date. The abundance of timber led to its use for construction, and this type of building was naturally very liable to be destroyed by fire in a country as unsettled as Worcestershire was until the seventeenth century. There is, however, one church, that at Ribbesford, where some of the pillars of the nave are of timber.

Ribbesford Church, showing wooden pillars

Many churches possess Norman doorways, and fine examples of architecture of that period may be found in the nave of Malvern Priory church and in the churches at Rock, Beoley, Stoke, Bromsgrove, Bredon, Chad-desley-Corbett, Holt, and Ribbesford. A splendid Norman font is to be seen at Chaddesley-Corbett; while at Ribbesford a curiously carved Norman tympanum tells the legend of the hunter, Robin of Horsehill, who aiming at a fawn on the further side of the Severn, transfixed both it and a salmon that leapt in midstream with the same shaft.

The Early English style may be seen in many churches in the county.   The choir of Malvern Priory

Doorway—Ribbesford Church

and parts of Northfield, Bromsgrove, Kempsey, and Pershore all belong to this period.

The limits of the fourteenth century roughly determine the period of the Decorated style.   Many Worcestershire churches contain work of this date, but attention must be especially drawn to those at Elmley-Lovett, Northfield, King's Norton, Kidderminster, and All Saints', Evesham.

The last Gothic period—the Perpendicular—prevailed from the reign of Richard II until the Reformation. The fine clock tower at Evesham, the handiwork of the last abbot, is a good example of the scope of the style. Many of the towers—that of Malvern Priory is especially fine—were built at this date, while the greater part of King's Norton church shows the form it assumed in smaller buildings. The east window at Malvern is said to be amongst the finest specimens of Perpendicular work in the country.

19.   Architecture.   (b) Ecclesiastical— Religious Houses.

Our knowledge of the ecclesiastical architecture of Worcestershire would be very incomplete if we did not consider the religious houses that once existed in the county, but of which we can now see only the ruins. Before the Reformation England was dotted over with abbeys, monasteries, and other religious houses, which were often fine specimens of the architect's skill.   In many parts of our land there yet remain enough of these buildings, either entire or in ruins, to impress upon us their beauty, and to convince us of the money and skill lavished on their construction.

In many instances the religious houses were centres

of beneficial influence ; they were the chief seats of learning, and they were the homes of men and women who helped the peasants, and promoted the prosperity of the country-side.   Henry VIII with ruthless hand, and guided by such advisers as Thomas Cromwell, decided to close the lesser monasteries first, and the greater monasteries afterwards.   Attached to the religious houses

Malvern Priory Church—Nave and East Window

were hundreds, and in some cases thousands of acres of the best land ; and the rent from these lands, together with other riches, helped to make the abbeys and monasteries very wealthy.   It was no doubt with a view to enriching himself and his administration that Henry VIII laid violent hands on the religious houses of England.   From his point of view he was successful ; but his harsh action told very heavily on the poor throughout the country, and it is now generally felt that Henry was not moved by a spirit of piety, but rather by a love of aggrandisement in thus suppressing the monasteries.

Now Worcestershire was a county in which Church influence was extraordinarily potent, and the monasteries must have represented before the Reformation a very important factor in it. We may be sure that they were civilising agencies of no mean order, and that they conveyed to the great mass of the people some of the best elements in their social and religious life.

The larger religious houses were at Worcester, Evesham, Pershore, and Malvern, while there were a crowd of smaller priories, such as those at Dudley, Halesowen, Dodford, Bordesley, Stoke, and Studley. When the Reformation came the change of proprietorship was not felt so extremely here as it might have been; for the bishop of Worcester kept his possessions, and the dean and chapter took over those of Pershore and of Malvern which had been a cell of Westminster. The latter included Malvern Priory and part of the lands of Pershore Abbey. But most of the churches and nearly all the monastic buildings fell into disrepair, and there is now very little left to indicate their former splendour and glory.

At Worcester there are the cloisters, chapter-house, and Edgar's tower, all of which formed part of the monastic buildings. Edgar's tower now serves as a muniment room for the early records of the monastery. It is a solid square tower over the gateway and was built in 1202.

Evesham Abbey was founded by Ecgwin, bishop of Worcester, about the year 700. Its possessions were co-extensive with the Hundred of Blackenhurst, and its magnificence must have been great ; but no traces of the church remain, and of the monastic buildings only the fine clock tower (built by Lichfield, the last abbot, in the

Edgar's Tower, Worcester

sixteenth century), the old gatehouse, almonry, cloister gateway and other fragments have survived.

It is pitiful to continue the account of the spoliation of the various other religious houses. Of the Abbey Church at Pershore, which once was 400 feet long, the choir and transepts alone remain, and at Malvern no traces of the monastic buildings, except the Tudor gatehouse, have come down to us, though here the greater

Abbot Lichfield's Bell Tower, Evesham

part of the church has been preserved. Some of the priories have been converted into farm-houses, but the majority have been demolished.

One small establishment, however, is worthy of notice. This is the Commandery or Hospital of St Wulstan at Worcester, which was a house belonging to the monks of the Augustinian order. The greater portion of it still

The Priory Gatehouse, Malvern

exists, including the large dining hall, sadly mutilated but with the minstrel's gallery at the one end and a little window at the other, from which a monk read the Scriptures during meals. The lands and revenues of this house, which was founded by St Wulstan, form part of the endowment of Christ Church, Oxford.

One other type of monastic building which must be noticed is the Tithe barn, in which the tithes due to the church (then paid in kind) were stored.   Examples of these, built of stone and of quite an ecclesiastical appearance, may be seen at Bredon and Middle Littleton.   In

Tithe Barn, Bredon

the particularly fine example at Bredon there is a small room over the porch in which a watchman might live, and the entrance is large enough to admit a loaded hay-wain.

20.   Architecture. (c) Military—Castles.

Castles are among the most conspicuous legacies be-
queathed to us by Feudalism, and their number bears
testimony to the grasp that the system had on the
country. Although not a few castles existed before the
Conquest, their real introduction into England was
brought about by the Normans.   After the Conquest
the lands were confiscated from the English and given
to the barons as a reward for their services.   To illustrate
the completeness of this change, we have only to recall
that it is estimated that 1100 castles were built at this
period.

It may be interesting to consider some of the general
details connected with these buildings. A castle of the
best construction consisted of a lofty and very thick wall,
with towers and bastions, enclosing several acres, and
protected by a moat or ditch.   Within this area were
three principal divisions.   First, there was the outer
bailey, or courtyard, the approach to which was guarded
by a towered gateway, with a drawbridge and portcullis.
In this bailey were the stables and a mount of command
and of execution.

Secondly, there was the inner bailey, or quadrangle,
also defended by gateway and towers. Within this second
court stood the keep, the chapel, and the barracks. Thirdly,
there was the donjon or keep, which was the real citadel,
and always provided with a well.

Some of the castles were royal ones, but every great
lord had one or two.   In many cases the position of

Norman castles was determined by considerations of strategy, and one would expect to find the passages of the Severn in Worcestershire guarded by them. Only a few castles are known to have existed in the county, and two of these guarded fords, one at Worcester and the other at Hanley Castle. A third was the bishop's residence at Hartlebury, while there were others, of which

Hanley Castle Church

little is known, at Elmley Castle and Castle Morton. Dudley Castle was in Worcestershire before the Conquest, but later came to form, as it does to this day, part of Staffordshire.

The scarceness of castles in Worcestershire is the more remarkable when we remember the troublous times that the county has gone through. The reason appears to have been that both Wulstan, bishop of Worcester, and the abbot of Evesham, who were the chief landowners in the county, gave their allegiance to the Conqueror and remained loyal throughout the first two revolts of the barons. Consequently although the sheriff, Urso d'Abitot, in building the castle at Worcester encroached on the precincts of the monastery, and brought down on himself Wulstan's well known curse,

> "Hightest thou Urse
> Have thou God's curse,"

the Church was nevertheless allowed to retain most of its possessions.

The castle at Worcester was on a large mound or *Buhr*, which had been thrown up in the time of King Alfred, as we know from a charter. It is not certain when the castle was demolished, but the mound was finally razed at the end of the eighteenth century.

Hartlebury Castle was burnt and nearly destroyed in Commonwealth times, and of Hanley and Elmley Castles nothing remains. Consequently the student of military architecture must go elsewhere; for Worcestershire at the present day has nothing to offer him.

21. Architecture. (*d*) Domestic— Famous Seats, Manor Houses, Cottages.

As in the military, so also in the domestic architecture of the county we find that the power of the Church has had its influence in reducing the number of large landowners and accordingly of large houses. Again, many small owners were brought into being by the constant attainders of the barons and especially by that of the Earl of Warwick in 1498, for the Crown claimed the property of the condemned man and re-distributed it in smaller holdings to less influential subjects. The Re-

A Tudor House, Broadway

formation also saw the introduction of several important Worcestershire families whose houses are of special interest when we are considering the domestic architecture.

But unfortunately the active part which was taken by the county in the Civil Wars brought about the destruction of many of the old houses. Among these we may mention old Westwood House, Frankley, and Hartlebury

Castle, just mentioned, the country seat of the bishop of Worcester, which were all burnt during the first war.

Within the limits of the county we may study the effect which the nature of the material at hand exercises on the architectural type. In the south-eastern part the once prosperous wool-manufacturing town of Broadway

Birtsmorton Court

with its large stone houses and grey slate roofs recalls the Cotswold manor-houses, of which a few may be seen in the county and its detached portions. Elsewhere stone is not much used for domestic buildings, and red brick or half-timbered houses predominate.

Among the larger residences we may mention Croome Court, Witley Court, Westwood House, Hewell Grange,

and Hartlebury Castle. The present houses are in each
case more or less modern.

During the time of the Tudor kings, men ceased
to build themselves strong castles, and many of the smaller
country seats in Worcestershire date from this period.
The most common type is the half-timbered manor-house
and of this there are many excellent examples. In these

Friar Street, Worcester

constructions a framework of wood was employed into
which bricks were built. Thus an economy in bricks
was obtained.

Some of the manor-houses are still surrounded by their
moats while others do not appear ever to have had any
such defence. Examples of the former may be seen at
Harvington and Birtsmorton. The latter affords an

excellent idea of what such a house was like in Stuart times since it still contains much of the original woodwork and furniture. As the Forest Court of Malvern Chase was held here, one large apartment formed the court room while there was a great banqueting hall on the other side of the small courtyard.

Unmoated manor-houses in very perfect preservation

The Old Almonry, Evesham

may be seen at Salwarpe Court and Wychbold Hall. These merge into the farm-house class which is here of course a large one. Many are half-timbered but a square-built red brick house with a low-pitched roof is the commonest type.

Though Worcestershire cannot boast any street so glorious as Tewkesbury High Street, yet Evesham,

Pershore, Droitwich, and Worcester have many fine old houses which are for the most part half-timbered. Bewdley has much good domestic architecture in its main streets ; but here, though wood is to some extent employed, the greater part of the construction is brick. The result is an appearance of greater solidity, attained, however, with a loss of picturesqueness.

It is perhaps when we reach such villages as Chaddesley Corbett, Ombersley, Cropthorne, or Feckenham, with their beautiful thatched cottages and neat though often garish flower-plots, that we see the type of architecture which one associates most closely with this county. These are but a few examples out of many and we may claim that almost any cottage in the agricultural districts makes an artistic picture and a pleasant contrast to the modern villa residence and jerry-built tenements which jostle one another in the manufacturing centres.

22. Communications. (a) Roads.

In order to understand the present arrangement of roads, we must glance at the history of their evolution. Starting as mere tracks from settlement to settlement, from camp to camp or from village to ford, the present system of highways has been gradually matured. At first there were the Ridgeways, then Saltways and Roman roads, and by direct descent from these the actual position and direction of our thoroughfares have been in many cases determined.

Archaeologists have pointed out that signs of Ridgeways

may be found in Worcestershire connecting the great earthworks of Malvern, Berrow, and Woodbury, while others can be traced running to one or two of the Severn fords.

The next most ancient set of roads in this county are the Saltways. Several of these roads, along which it is supposed that Droitwich salt was carried, are claimed to be still traceable. The best marked is the Upper Saltway, which ran in a north-easterly direction from Droitwich through Bromsgrove and Birmingham, and so to Lincoln. The Lower Saltway or Sale-way led from Droitwich to Alcester. It is probable that these Saltways existed in pre-Roman times, but as they now appear they are most probably of Roman work.

In order to understand the Roman roads of this district, we must remember that the county as such did not then exist. The region lay roughly between four trunk roads;—on the south-east Icknield Street and the Fosse Way, on the north-east Watling Street, on the west the road from Hereford to Chester, and on the south-west that from Cirencester to Hereford.

Of these the Icknield Street runs through the eastern side of the county, and so we may expect to find cross roads connecting with it from the west. The road itself enters the county in the south as the Buccle Street, reaches Honeybourne and runs thence almost due north through Alcester, Beoley, and King's Norton. Beyond this it is joined by the Upper Saltway and travels in the same direction as before to Wall near Lichfield. The Lower Saltway crosses it at Alcester.

w. w.   8

The great Fosse Way, which connected Devon and Lincoln, runs through some of the detached portions of Worcestershire and we must not forget that Icknield Street is really a branch of this road. It was by these two ways that the traveller from the north-east could reach the great Roman centre of Corinium (Cirencester).

Besides the two Saltways, it is probable that a Roman road ran up the east side of the Severn, through Kempsey

Powick Bridge

(where there has been found a Roman monument commemorating the Emperor Constantine) to Worcester, where it forked, one branch going to Droitwich while the other followed the line of the Severn in a northerly direction into Staffordshire. The existence of the latter road is however not yet thoroughly proved.

It is curious how the old names of some of these ways

have survived. For example we have in Worcestershire the Sale-way, the Portstreet at Bengeworth, Portways at Beoley and other places, and Silver Street (*silva*, a wood) in Worcester itself.

In the Middle Ages it is certain that Worcester became one of the great commercial centres through which the trade with Wales passed, but the exact direction that the main part of the traffic took is not determinable. Probably, however, the route along the Teme valley was the most important.

When we speak of roads before the beginning of the nineteenth century we must not imagine highways such as we are accustomed to now, but must picture to ourselves a track full of dust in dry, and of mud in wet weather, uneven and unpaved except in the towns and then only with cobble stones. We read that about the year 1800 people in the Evesham and Pershore district formed a society for improving the roads and in 1810 Pitt writes that "60 to 80 horses have been formerly laden in a day with garden stuff for Birmingham but the roads being now improved, it is sent in wheeled carriages." But this is only a sign of the times. For with the discovery by McAdam of the use of broken stones for roadmaking the quality of the roads improved everywhere, and by about 1820 they were in first class order. It was at this time that coaching was at its highest pitch of perfection. We find that the journey from London to Worcester was timed to take 14 hours, which is remarkably rapid travelling when we remember that the distance by road between the two places is 111 miles.

At the present day the county contains many first-class high roads, while the quality of the by-roads is distinctly good. There is an abundance of hard stone suitable for road-metal and this is no doubt an important factor in determining the quality of the roads. At the same time it may be noticed that imported stone is very frequently used for the main roads. The Worcestershire lanes are among the most beautiful in England and now that the highways are beset with motor cars and the consequent dust, pedestrians and cyclists are led more and more to use these winding routes.

23. Communications. (*b*) Canals and Railways.

The vast majority of our English canals were built in the period between 1750 and the introduction of railways in 1837. At that time they were far more used than at present, and the facilities for travelling and transporting goods by them were better. Consequently this period is sometimes spoken of as the Canal Era.

It was the railways that brought about the decay of the canal system; for many of the latter were bought up or "comptrolled" by the Railway Companies in order to avoid competition. Canal transport, compared with that by rail, is very slow, and accordingly is only employed for a very limited variety of goods, such as coal, raw materials, bricks, and heavy or bulky manu-factured articles. Further, a want of standardisation and of unity of control, has led to the extinction of through-traffic on the longer routes. If these conditions could be modified and our canals be brought up to date as regards locks, as the recent Royal Commission thinks possible, there seems no reason why our inland waterways should not enjoy prosperity again.

Such improvements would especially affect the Midlands, with its large manufacturing centres and its want of outlet to the sea. Two of the existing waterways in Worcestershire are considered to be capable of development into first-class through-routes. These are the Staffordshire and Worcestershire Canal, and the Worcester and Birmingham Canal, both of course in conjunction with the Severn.

We may gather some idea of the importance that canals formerly possessed by an examination of those which are found in Worcestershire.

In the first place, the Severn was once far more extensively used than now, and we know that throughout the Middle Ages there were continual quarrels between the bargemen and the city authorities at Worcester, who claimed a toll from every boat that passed beneath the bridge. However, it finally became recognised that the river was the king's high waterway, an Act of 1430 reciting that "the river Severn is common to all the king's liege people to carry and to recarry all manner of merchandise as well in trowes and boats as in flotes otherwise called drags."

The current even now, when locks ensure 10 feet of water all the way up to Stourport, is very strong and is probably the chief drawback to navigation on this river. Barges are now roped one to another in a long string and towed up and down by small tugs.

The river Avon was once navigable right up to Stratford, but now the locks are mostly out of repair and no trade is carried on by its means.

The earliest canal projected in Worcestershire was one to connect the Severn and the Trent. This canal, originally known as the Trent and Severn Canal, but now called the Staffordshire and Worcestershire Canal, was opened about 1755 and it was this which brought the town of Stourport into existence at the point where the canal joined the Severn. We know that at the beginning of the nineteenth century this route was much used for transporting fruit, as much as 7000 tons going over it in a year when there was a "hit" of fruit in Worcestershire. A branch of this canal reaches Stourbridge. These routes were, however, and still are, chiefly occupied with iron, coal, and grain traffic to and from the Midlands.

The Droitwich Canal was cut in 1771. This is still used by barges and sailing wherries. At the time of its building it was intended as a means of bringing coal to Droitwich. An ingenious scheme to avoid the coal difficulty was unsuccessfully exploited by a London druggist, by name Baker, who proposed to carry the brine in pipes to the Severn, where coal could be more readily obtained.

The canal which was the most difficult to engineer was the Birmingham and Worcester Canal. The promotors were met by the same difficulty which the railway engineers experienced 70 years later, in that Birmingham lies about 500 feet above sea-level on high ground dropping rather steeply to the plain of Worcestershire. To overcome this difference of level a string of locks some 30 in number had to be built near Bromsgrove. The time spent in passing these is a serious disadvantage to the route, and at present the canal is not much used.

Lastly there is the Birmingham and Dudley Canal, which is still of importance since it is mainly occupied in the transport of coal, iron, bricks, and heavy goods, in which quick transit is not imperative. Its traffic is entirely local.

A comparison of these two canals is of interest ; for they exemplify a very general principle in English Canal enterprise at the present day. The Birmingham and Dudley Canal with its purely local traffic is exceedingly prosperous, while an important through-route, like the Worcester and Birmingham Canal, cannot compete with the railways, even for heavy raw materials and coal.

The chief railway artery of Worcestershire is the Midland Railway Company's main line from Birmingham via Worcester and Gloucester to Bristol. The most important branch line from it is the loop through Redditch and Evesham, and back to the main line at Ashchurch Junction. There is also a short branch from Northfield to Halesowen. The main line is notable on account of the very steep gradient between Bromsgrove and Blackwell, which is as much as 1 in 37. Ascending trains need the assistance of an engine behind, and it is sometimes possible to see as many as four engines to one train.

The other company serving the county is the Great Western Railway Company whose line runs from Wolverhampton, through Stourbridge and Droitwich, to Worcester. Here it branches, one line going to Malvern, Ledbury, and Hereford, and the other through Evesham and Honeybourne to Oxford and London.

It will be seen from this account of the railway communications that there is every facility for that quick transport of fruit, vegetables, and farm produce to good markets, which is so essential for an agricultural district.

24. Administration and Divisions. Ancient and Modern.

Before we consider the present administration of affairs in the county of Worcester, it will be well for us to get some idea of the ancient forms of government. It must first of all be clearly understood that many changes have been introduced into our parochial and county government since the time of our Saxon forefathers ; but it has been the great care of most of our rulers to graft, as it were, new ideas on to the old English institutions. Thus it comes about that, although we have improved methods of government to suit modern ideas, we can trace back many of our present institutions for a thousand years or more.

The government of a county or shire in early English times was partly central, from the county town, and partly local, from the hundred or parish. The chief court of the county of Worcester was, in early times, the Shire-moot, which met twice a year. Its principal officers were the Ealdorman and the Sheriff, the latter of whom was appointed by the King. In Saxon times, each county was divided into Hundreds, or Lathes, or Wapentakes. Worcestershire was divided into twelve Hundreds, and it is probable that, at first, each of these divisions contained one hundred free families. Each Hundred had its own court, the Hundred-moot, which met every month for business. At the present day there are eleven Hundreds in Worcestershire.

Each Hundred was divided into townships, or parishes as we now call them. Each township had its own assembly or *gemot*, where every freeman could appear. This gemot, or town-moot, made laws for the township, and appointed officers to enforce their *by*-laws, or laws of the town. The officers of the town-moot were the "reeve" and the "tithing-man," who was the constable and corresponded to our policeman. The court of the township was held whenever necessary, and the reeve was the president or chairman.

Besides these courts of the shire, the hundred, and the township, there were also courts of the Manor, as the separate holdings of land were called. The manors were of different sizes; sometimes they were as large as the township, and sometimes they were parts of the township. The manors were held by their owners, or lords of the manor, as they were called, on various conditions. For example, they had to render service or homage to the King, and were allowed to sub-let their manors. The manor-courts were various, and were called Court-leet,

Court-baron, and Customary court. In these courts, the lord and his tenants met, and dealt with cases occurring in the manor, such as those relating to the common fields, the rights of enclosure, and the holding of fairs and markets, and also with criminal offences.

We are now in a position to consider the present form of administration of public affairs in the county of Worcester. The chief officers in the county are the Lord Lieutenant and the High Sheriff. The former is generally a nobleman, or a large landowner, and is appointed by the Crown. The Sheriff is chosen every year on the morrow of St Martin's Day, November 12.

The County Council now conducts the chief business of the county of Worcester and has its headquarters at that city. It consists of 21 aldermen and 63 councillors. The latter are elected to their office whereas the former are co-opted by the councillors for a term of years. The County Council corresponds to the ancient Shire-moot, and represents the central form of county government which was started in 1888.

For local government in towns and parishes another Act was passed in 1894, when new names were given to the local governing bodies which had previously been known as vestries, local boards, highway boards, etc. In the large parishes, the chief governing bodies are now called District Councils, and of these there are 15 in Worcestershire. The smaller parishes have Parish Councils, or Parish Meetings. But whether District Councils or Parish Councils, they represent the old Town-moots, and are chosen by the votes of the people in the parish.

Their members are elected to manage the local affairs of the place, and generally to advance its interests.

There are some towns in Worcestershire that have larger and different powers of government than the parishes. These towns are called boroughs and are as follows :—Bewdley, Droitwich, Dudley, Evesham, Kidderminster, and Worcester. Worcester and Dudley are the largest and most important, and have the dignity of county boroughs with the power of County Councils. Worcester is also a county to itself, having been so constituted by James I.

Worcestershire has 13 Poor Law Unions, each of which has a Board of Guardians, whose duty it is to manage the workhouses, and to appoint relieving officers and others to carry on the work of keeping the poor and the aged.

For purposes of administering justice, a Court of Quarter Sessions is held both at Worcester and at Dudley. There are, further, 19 Petty Sessional Divisions, each having magistrates or justices of the peace, whose duty it is to try cases and punish petty offences against the law. The other boroughs have Commissions of the Peace only.

In early times Worcestershire suffered not a little from two peculiar courts—the Court of the Welsh Marches, whose decrees were often a scourge rather than a blessing, and the despotic Forest Courts. Large parts of the county, being forest land, came under the jurisdiction of the latter. At Birtsmorton a Court House of Malvern Chase is still standing.

Birmingham University, Selly Oak

If we go back to the earliest days of the Saxon Conquest, we find that the Church as a body existed before the State. Its mode of government is much the same to-day as it was in those far-off times. Our land was divided into dioceses, over which were placed bishops. The northern dioceses and bishops were under the care of the Archbishop of York ; while the southern dioceses and bishops were under that of the Archbishop of Canterbury. Drawing its first bishops from Whitby in Yorkshire, Worcester came at first under York, but it was soon transferred to Canterbury.

The diocese of Worcester, originally covering Gloucestershire, Worcestershire, and Warwickshire, was curtailed by Henry VIII, who separated Gloucestershire as a new diocese. It has been further diminished by the creation of a diocese of Birmingham. The whole of Worcestershire is not in the diocese of Worcester, for 22 parishes belong to the see of Hereford, and others to that of Birmingham.

Each diocese is divided into archdeaconries, which in turn are subdivided into rural deaneries and ecclesiastical parishes. At one time the ecclesiastical parish was the same as the civil parish, but now while there are 239 of the latter in Worcestershire, there are many more ecclesiastical parishes.

The educational affairs of Worcestershire are managed for the most part by an Education Committee appointed by the County Council, though Kidderminster, Oldbury, Northfield, Dudley, and Worcester manage their own elementary education.

Worcestershire is represented in the House of Commons by eight members of parliament. The following three parliamentary boroughs each send one member :—Worcester, Dudley, and Kidderminster; the rest of the county is divided into five divisions for parliamentary representation, and each of these sends one member.

25. Roll of Honour.

Worcestershire has its quota of notable names connected with it by ties either of birth or association. Though no kings or queens have been born in the county, yet King John, Prince Arthur, and Charles II are intimately associated with its history. Many of the earlier kings, and also Queen Elizabeth, visited Worcester, and hunted in the forests of Feckenham, Wyre, and Malvern, and there was a royal palace at Ticknall near Bewdley.

King John appears to have been especially fond of Worcester, and in accordance with his will, was buried in the Cathedral between the tombs of St Wulstan and St Oswald. Not far from these graves is the glorious Chantry chapel raised by Henry VII in memory of Prince Arthur, who was married by proxy at Bewdley. He died at Ludlow in 1502 and was buried in Worcester.

The memory of Charles II and his connection with Worcester must not be passed over. The "loyal city" suffered much on his behalf and is to this day proud of its wounds and of the defeated monarch whose cause it so strenuously upheld.

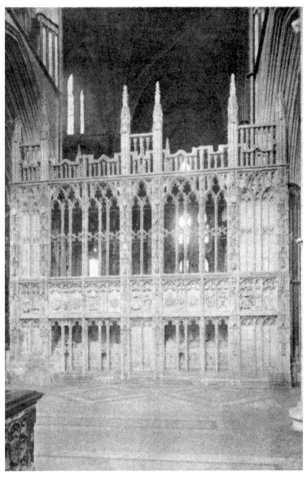

Prince Arthur's Chantry Chapel, Worcester Cathedral

When we come to consider the statesmen that the county has produced we find one or two famous names. The first of these is Lord Somers, first Baron of Evesham, who was born near or in Worcester in 1651.

John, 1st Lord Somers

He is best known for his share in the defence of the Seven Bishops in 1688 and as Lord Chancellor of England. He was also a friend and patron of a number of literary men including Addison and Steele.

The first Lord Lyttelton was also a Worcestershire

man. He was at one time Chancellor of the Exchequer and, like Lord Somers, was a patron of literature. He himself wrote a number of books, of which *Dialogues of the Dead* and the *History of the Life of Henry II* are the best known.

Warren Hastings

Next, we must include in our roll of honour Warren Hastings, who was born in 1732 at Daylesford, one of the detached parts of Worcestershire surrounded by Gloucestershire and Oxfordshire. This famous governor-general must be regarded as one of the chief founders of our

Indian Empire. His trial on the charge of cruelty and corruption during his rule in India is one of the longest on record, lasting as it did from 1788–1795, when he was finally acquitted.

Worcester may lay claim to a number of celebrated bishops and divines. The first really great bishop of Worcester was St Oswald, whose introduction of the Benedictine rule into the Worcestershire monasteries had such beneficial effects on the civilisation of the inhabitants of the county. Then we must recall Wulstan, the great Englishman who stood by the Conqueror and his successor when their rule was threatened by the revolts of the barons. He it was who built the crypt of the present Cathedral and founded the Commandery. Later we find Walter de Cantilupe, more of a statesman than a bishop, who was one of the chief defenders of English liberties against the Crown and Papacy during the reign of Henry III. He helped Simon de Montfort to cross the Severn the day before the battle of Evesham, where Simon was slain. The latter after his death was revered as a saint, and his tomb became celebrated for the miracles wrought at it and drew many pilgrims to Evesham Abbey.

Hugh Latimer, who was martyred at Oxford in 1555, was bishop of Worcester; and we may here mention the celebrated nonconformist divine, Richard Baxter, who lived most of his life at Kidderminster, where there is a monument to his memory.

The literary names connected with Worcestershire are more remarkable as showing the state of culture attained in this district in the middle ages than for anything else. Thus we may lay claim to two of the earliest English poets; Layamon (*c.* 1200), a priest at Arley Kings, whose poem in English, *The Brut*, purported to be a history of early England, and William or Robert Langland (*c.* 1330),

Samuel Butler

to whom is ascribed *The Vision of Piers Plowman.* This work appears to have been addressed to the people and to have been exceedingly popular, attacking as it did the misrule of Richard II. It is possible that more writers than one were concerned in the work.

Again, there was Florence of Worcester who wrote about 1110. His work *Chronicon ex Chronicis* was a general history from the beginning of the world to the last years of his life.

Of later authors we must mention Samuel Butler (1612–80), who lived at Strensham and is best known by his poem *Hudibras*; and Shenstone (1714–63), who was a contemporary of Dr Johnson and who wrote *The School-mistress*. He devoted most of his attention and fortune to beautifying his house at the Leasowes near Halesowen.

There are two historians of the county. Habington was condemned to death for complicity in the Gunpowder Plot but afterwards pardoned on condition that he should not leave the county. Thus confined to Worcestershire he set himself the task of writing its history, which was largely drawn upon by Nash, whose *Collections for a History of Worcestershire*, published about 1780, is somewhat of a classic. Nash was born at Kempsey, and Habington passed most of his days at Hindlip near Droitwich. The former was for Worcestershire what Dugdale was for Warwickshire.

Rowland Hill, who introduced the present system of penny postage and suggested the use of adhesive stamps, was born at Kidderminster in 1795.

Lastly, the county may claim among her illustrious men one of the finest printers of modern times—John Baskerville, who was born at Wolverley in 1706.

26. THE CHIEF TOWNS AND VILLAGES OF WORCESTERSHIRE.

(The figures in brackets after each name give the population in 1901, and those at the end of the sections give the references to the text.)

Acocks Green (3836). A suburb of Birmingham. At Tyseley there is a junction on the Great Western Railway, a new line running from here to Henley-in-Arden.

Alvechurch (1731). A picturesque and ancient village on the river Arrow. The church has been rebuilt. Formerly there was a summer residence of the bishop of Worcester here. There are needle mills in the parish. Mummers perform an old play here at Christmas, with ancient doggerel rhymes. (p. 72.)

Belbroughton (1851). A picturesque village with ancient parish church. The brook provides water-power to several factories where scythes, hay and chaff knives and other edge tools are made. (p. 61.)

Beoley (565). A little village just off the Icknield Street. The name is a corruption of Beau lieu. The church contains a very fine early Norman font, the major part of the building being in the Early English and Perpendicular styles. (pp. 96, 115.)

Bewdley (2866). An old town on the Severn with a bridge built by Telford which replaced an older one that had been destroyed by floods. Formerly a great centre of the woollen trade, was later noted for its manufacture of horn articles. In Elizabethan times it manufactured cloth caps. The parish church is at Ribbesford and has the pillars on one side of the nave made of wood and a fine Norman tympanum over the door. Tickenhill, or Ticknall Palace, used to stand near Bewdley. It was a royal palace and it was here that Prince Arthur was married by proxy to Catherine of Aragon in 1499. Wyre Forest is to the north-west of the town. The river is used for boating and the place is quite a holiday resort for people from the Black Country and Kidderminster. (pp. 7, 19, 20, 60, 64, 80, 96, 123.)

Blockley (1812). A detached parish with an old church. Northwick Park did contain a fine collection of pictures. (p. 12.)

Bournville. A suburb of Birmingham. There is a model village here in connection with Cadbury's cocoa and chocolate works. (p. 65.)

Bredon (1070). A picturesque agricultural village below Bredon Hill. The twelfth century church is of interest and contains good Norman work and a fine monument to Giles Reed. There is also a noteworthy tithe barn. (pp. 14, 16, 96, 104, 135.)

Broadway (1414). A stone-built village at the foot of the steep Cotswold escarpment. It is much resorted to by artists and motorists. Formerly a centre of the woollen industry. (p. 109.)

Bromsgrove (14,106) (Saxon *Bremesgrave*). An old market-town with an Edward VI grammar school and ancient church. Formerly a great centre of the nail-making trade which has however declined. There are railway-wagon works, boot and button manufactures, sandstone quarries, and a flourishing guild of applied arts in the town. In the neighbourhood, at Stoke, are large saltworks. (pp. 43, 59, 61, 62, 64, 96, 97, 119.)

Catshill (2785). Formerly part of Bromsgrove. Its chief industries are nail-making and market-gardening.

Chaddesley Corbett (1392).   The church contains a fine Norman font and is the only one in England dedicated to St Cassian.   (pp. 96, 112, 136.)

Church Honeybourne (81) is an important junction on the Great Western line, branches going to Cheltenham and to Wercester.   (pp. 113, 120.)

Bredon Church

Crabbs Cross and **Astwood Bank** (1630) is a part of Feckenham parish.   Sewing-machine needles are made here.

Cradley (6733).   A large centre of the chain-making industry.   There are also firebrick, clay retort and crucible manufactures, and iron works.   The church is modern.   (pp. 61, 70, 73.)

Cropthorne (356).   A picturesque village given up to market-gardening.   The church contains fine monuments to the Dineley family.   (pp. 89, 112.)

Font in Chaddesley Corbett Church

Droitwich (4164).   One of the oldest towns in the county, there having been a Roman settlement here.   From time immemorial it has been noted for its salt.   This industry is still extensively carried on, and the brine is also successfully used as baths for the cure of rheumatism.   Consequently Droitwich has become a spa of no small importance.   The churches of St Andrew and St Peter are both ancient.   The town is served by two rail-

Hampton Ferry, Evesham

ways and two canals.   Westwood House is a large residence in the neighbourhood.   (pp. 35, 60, 66—69, 72, 73, 88, 111—113, 118, 120, 123.)

Dudley (48,733).   This populous town is the centre of various hardware trades and is of ancient foundation.   The Norman castle lies in Staffordshire.   The town is served by trams, railways, and canals.   The presence of limestone, coal, and ironstone in the neighbourhood has caused its importance. The following are among the manufactures of Dudley:—iron and brass founding, glass, nails, fenders, anvils, boilers, anchors, chains, grates, railway bridges and wagons, iron and steel constructions, and edge tools. (pp. 12, 33, 51, 61, 70, 71, 73, 82, 123, 126.)

The Market Place, Evesham

Evesham (7101). An ancient town which arose round the abbey. This has been entirely demolished, except the Bell-tower and a few isolated buildings. The bloody battle in which Simon de Montfort, 160 knights, and about 4000 men were slain and the barons' cause overthrown was fought on Green Hill in 1265. The town was also the scene of various encounters in the Civil War and was stormed in 1645 by the Parliamentary army under Massey. There is a grammar school founded by Abbot Lichfield in Henry VIII's reign.

At the present day Evesham is a busy country market-town with market-gardening and fruit-growing as a staple industry. Asparagus is sent all over England from here. Of late the intensive mode of culture of vegetables and fruit, so successful in France, has been tried with success here.   (pp. 8, 23, 35, 53, 59, 80, 98, 100, 101, 111, 115, 123, 130.)

Feckenham (3925).   A picturesque village on the Lower Saltway about halfway between Droitwich and Alcester.   It was from this village that the Forest of Feckenham took its name. (pp. 18, 68, 112.)

Fladbury (1172) was known in Saxon times as Fleodaubyrig. It lies on the Avon about four miles below Evesham and is a popular resort by steamer and boat from that town.

Hagley (1399) lies on the west side of the Clent Hills about two and a half miles from Stourbridge.   Hagley Hall has been the seat of the Lyttelton family for centuries.   (p. 82.)

Halesowen (10,550) is situated four miles east of Stourbridge and is a hardware centre with an old church dating from Norman times.   The following articles are manufactured here:— gunbarrels, anchors, anvils, nails, rivets, gas tubing, edge tools, files, and horn buttons.

In pre-Reformation times there was an abbey here of Premonstratensian monks, but practically nothing remains of it now. (pp. 100, 119, 132.)

Hanley Castle (1098) lies about nine miles south of Worcester. No remains of any castle survive now. (pp. 106, 107.)

Hartlebury (2442).   A small village lying near Hartlebury Castle, the seat of the bishop of Worcester.   The old fortified house was nearly destroyed in the Civil Wars and the present building dates from Charles II's time.   There is an interesting collection of portraits of bishops.   Part of the house is now used as a clergy college.

Monument in Kempsey Church to Sir Edmund Wylde

The village church has been mostly rebuilt. There is a grammar school founded by Richard III. The " Mitre Oak " is one of the famous trees of the county. (pp. 41, 82, 106—108.)

Inkberrow (1461). A picturesque village, five miles west of Alcester, with an old church. Sandstone is quarried here.

Kempsey (1518). This village is on the site of a Roman camp four miles south of Worcester. A stone has been dug up here commemorating Constantine the Great. The church is in the Early English style. (pp. 80, 97, 114, 132.)

Kidderminster (27,745) lies on the river Stour. There is a fine church in the Early English, Decorated, and Perpendicular styles. The town is served by the Great Western Railway, by trams and by the Stourbridge canal.

The industries are carpet-making, worsted-spinning, and dyeing. Richard Baxter, the famous nonconformist divine, spent most of his days here. (pp. 38, 63, 64, 98, 123, 126, 130, 132.)

Kingsheath (10,078). A residential suburb of Birmingham.

King's Norton (12,227). An outlying suburb of Birmingham some five miles from the centre of the town. It has a fine church with an old half-timbered schoolhouse in the churchyard. Hawkesley Hall, a mile to the south, was garrisoned for the Parliament during the Civil War. (pp. 19, 98.)

The Lickey (3705). A straggling residential and agricultural district, north of Bromsgrove and on the sides of the Lickey Hills. The latter are much frequented as an open space belonging to Birmingham. (pp. 14, 30, 36, 38, 43, 46.)

Great Malvern (8669). This town nestles at the eastern foot of the Malvern hills, which are chiefly common land. It is a considerable health resort, having extraordinarily pure water. The Priory of Malvern was founded in 1083. The Priory

Church is a glorious building with a Norman nave and grand Perpendicular tower. Most of the monastic buildings have been destroyed. There are numerous schools in the district of which Malvern College is the best known. (pp. 30, 48, 72, 97, 98, 100, 101, 120.)

Malvern Wells (1559). A residential village on the slope of the hills south of Great Malvern. Several important schools have selected this locality. A celebrated Holy Well is here. A

Miserere Seat, Priory Church, Great Malvern.
Rats hanging Cat

mile south of the village is the picturesque Priory and Court of **Little Malvern.**

West Malvern (1406) is also a residential district on the northern and western flanks of the hills, commanding fine views into Shropshire, Herefordshire and Wales.

North Malvern. A large village north-west of Great Malvern. Its population is reckoned with the other Malverns.

Malvern Link Common

Malvern Link (4814) is a residential district about two miles north of Great Malvern.

Moseley (1174). A suburb of Birmingham lying about two miles to the south of the centre of the town.

Northfield (20,767). A large village six miles south-west of Birmingham, with an ancient church. It is chiefly a residential district. Motor cars are made here. (pp. 97, 98.)

Malvern College

Oldbury (21,467), with **Warley** (3724) is a hardware centre, with new churches. It is especially noted for its glassworks which manufacture the large lenses for lighthouses. Its other manufactures are iron, steel, edge tools, railway carriages, and chemicals. (p. 62.)

Ombersley (2017). A delightful village with a good but modern church, is situated five and a half miles north of Worcester. (p. 112.)

Pershore (2826) is the centre of the fruit-growing districts of the lower Avon valley. It lies on the Avon and has a fine abbey church, a remnant of the Abbey itself, whose nave was demolished at the Reformation. St Andrew's church was founded by Edward the Confessor for the tenants of Westminster Abbey, which latter owned much of Pershore. It is in the Norman style. (pp. 8, 23, 58, 80, 97, 100, 101, 112, 115.)

Redditch (9438) is a flourishing little town, given over to the manufacture of needles, fish-hooks, and fishing tackle.

The Avon at Pershore Bridge

Gramophone and sewing-machine needles, and bicycles are also made. The town is quite modern. At Bordesley there was formerly a Cistercian abbey, but there are no remains of it to be seen. (pp. 59, 63, 119.)

Ripple (685) is a picturesque village with a cross, whipping-post, and stocks in the street. It lies four miles north-west of Tewkesbury (p. 146).

The Cross, Stocks and Whipping Post, Ripple,
near Tewkesbury

Rock (1513). This is a scattered district including part of Wyre Forest. Cherries are grown here. The church is a good example of Norman work. (pp. 19, 96.)

Rous Lench (207) is one of the three "Lenches" lying north of Evesham. It has an interesting church with some Saxon and Norman work in it. (p. 95.)

Selly Oak (16,222). A manufacturing suburb of Birmingham lying about four miles to the south-west of the town. The new University Buildings are here. The manufacture of bicycles and bicycle-fittings, and metal-rolling are the chief industries. (p. 61.)

Shipston-on-Stour (1564) is a detached portion of Worcestershire. It is a small market-town, with an October fair at which a "Bull-roast" is held. The church is modern. (p. 12.)

Stourbridge (16,302) as it now stands is nearly all modern. It is the centre of the fire-clay and glass industries. There are also iron works and glue factories. Glazed bricks and porcelain baths are made here. The town is served by trams, railways, and canal. (pp. 49, 60, 62, 71, 118, 120.)

Stourport (3111). This is a modern town, situated where the Stour reaches the Severn. Here too the Trent and Severn Canal, now known as the Staffordshire and Worcestershire Canal, joins the latter river. Its importance was formerly bound up with these waterways, but at present its trade consists chiefly of carpet-making, ironworks, and the manufacture of hollow-ware and hinges. (pp. 20, 38, 60, 117, 118.)

Strensham (233). A small village, four miles north of Tewkesbury, famous as the home of the Russells and of Samuel Butler the author of *Hudibras*. There are monuments in the church to these families. (pp. 82, 132.)

Tenbury (2080) is situated in the extreme west of the county, on the river Teme. It is sometimes known as Tenbury Wells. The church has a Norman tower. The town is a small spa, with medicinal waters. There is good fishing in the Teme, and the surrounding country is very beautiful. The town is the centre of a large hop-growing district. (p. 21.)

Upton-on-Severn Old Church

Upton-on-Severn (2225) is a quiet village on the lower waters of the Severn. The bridge was the scene of the valiant deed of Fleetwood's 18 men who on 29th August, 1651, crossed on the single planks left by Charles's force. Later that day the battle at Upton was fought. The old bridge itself was destroyed by a flood in 1852. Ham Court in the neighbourhood is a fine half-timbered house.

Witley (488). Here the Earl of Dudley has a large residence, Witley Court, which is a house in the seventeenth century Italian style with magnificent grounds. It lies to the south-east of Abberley Hill.

Worcester (46,624) is a city, a county in itself, the see of a bishop, capital of the county, and a parliamentary and municipal borough. It is situated on rising ground above the Severn on its left bank; the river is spanned by a stone bridge. Formerly there was a castle here close to the cathedral which is well placed near the river and surrounded by various ecclesiastical buildings. The cathedral was founded by the Saxons, but the earliest surviving portion is the crypt built by St Wulstan. The Norman building was destroyed by fire, and accordingly most of the present church is later. The choir was begun in 1224 and is in the Early English style while most of the nave belongs to the Decorated Period.

The diocese formerly embraced Gloucestershire as well as Warwickshire and Worcestershire, but the former was made into a new see in 1541 by Henry VIII. Under the Conqueror there was a mint at Worcester and money was coined there as late as Charles I's reign, a pear being the mint mark.

There are 11 parish churches in Worcester of which St Helen's, St Alban's, and St Andrew's with the slender spire, are the most interesting. The Commandery, and King Charles' House in New Street, are specially worthy of note. The former was the Royalist head-quarters during the battle of Worcester in 1651. The Guildhall is a curious building of the time of Queen Anne.

The staple industries of Worcester are glove-making, and the manufacture of porcelain, vinegar, and railway signals. Berrow's *Worcester Journal* is the oldest British newspaper, having been started in 1690. There are various schools, of which we

The Commandery, Worcester

may note the King's School founded by Henry VIII out of the Monastery School when that institution was suppressed, the Cathedral and Choir School, and the Worcester Royal Grammar School. (pp. 3, 4, 12, 60, 62, 63, 65, 75, 77, 79, 80, 81, 88, 92—95, 100, 103, 106, 107, 115, 117, 120, 123, 126, 130.)

Yardley (33,946) is a large suburban district, four miles to the east of Birmingham. There are wire mills and tile factories here. (p. 72.)

Worcester Cathedral, Nave, looking East

England & Wales

37,130,344 acres

Worcester
480,560

Fig. 1. The Area of the Ancient Geographical County of Worcestershire (480,560), compared with that of England and Wales

England & Wales

Population 32,527,843

Worcester
488,338

Fig. 2. The Population of Worcestershire (488,338) compared with that of England and Wales (in 1901)

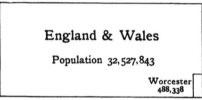

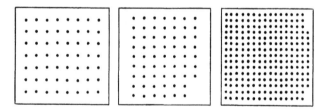

England and Wales 558   Worcestershire 605   Lancashire 2347

Fig. 3. Comparative Density of Population to Square Mile in 1901

(*Each dot represents ten persons*)

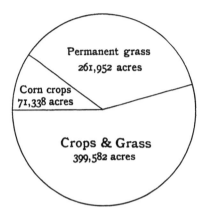

Fig. 4.   Proportion of Corn Crops to Pasture in
Worcestershire in 1907

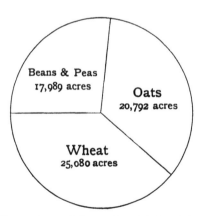

Fig. 5.   Proportions of the various kinds of Corn Crops
in 1907

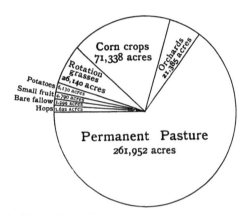

Fig. 6. Proportion of Permanent Pasture to other crops
in Worcestershire in 1907

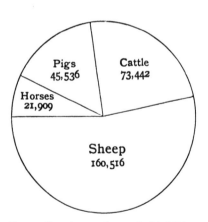

Fig. 7. Proportionate numbers of chief Live-stock in
Worcestershire in 1907